The Institute of Refrigeration Proceedings 2021-2022

Volume 118

2021-2022

ISBN: 978-1-7138-8037-0

Printed from e-media with permission by:

Curran Associates, Inc.
57 Morehouse Lane
Red Hook, NY 12571

Some format issues inherent in the e-media version may also appear in this print version.

The Institute of Refrigeration (IOR)
Kelvin House, 76 Mill Lane
Carshalton, Surrey SM5 2JR
Charitable Incorporated Organisation 1166869

Phone: +44 (0) 20 8647 7033

www.ior.org.uk

Additional copies of this publication are available from:

Curran Associates, Inc.
57 Morehouse Lane
Red Hook, NY 12571 USA
Phone: 845-758-0400
Fax: 845-758-2633
Email: curran@proceedings.com
Web: www.proceedings.com

IOR.org.uk
REFRIGERATION AIR CONDITIONING HEAT PUMPS

The knowledge hub for refrigeration, air conditioning and heat pumps

The Institute of Refrigeration

PROCEEDINGS 2021–22
Volume 118

The Institute of Refrigeration (IOR) is an independent registered charity run for the public benefit. Its membership is over 2000 individuals whose work depends on refrigeration, air conditioning and heat pumps, providing technical information for designers, specifiers, installers, end users and service and maintenance personnel.

IOR publishes a range of Refrigerant Codes of Practice, Guidance Notes, Good Practice Guides and Safety Alerts. The ever growing library of up to date and reference publications contributes to making the IOR a central point of reference and resource for people for all matters relating to refrigeration, air conditioning and heat pump practice. IOR also offers events, webinars and e-learning. Visit www.ior.org.uk/membership for further information about our activities and how you can get involved.

The Institute of Refrigeration

PROCEEDINGS
Volume 118
2021–22

CONTENTS

Minutes of Meetings of Members 2021-22

4th November 2021
A meeting of 41 members and guests of the Institute of Refrigeration was held at Park Crescent Conference Centre in London. The meeting was chaired by IOR President Graeme Maidment FInstR. The paper entitled "RACHP-The Journey to Net Zero" was presented by the authors Robert Lamb FInstR, Philip Holzer MInstR. Graeme Fox FInstR, Neil Hewitt MInstR. Tom Southall. Questions were asked by Mike Creamer FInstR, Martin Craxton, David Paget FInstR and Nick Frazen MInstR. The meeting followed the IOR AGM and closed at 6.40pm.

2nd December 2021
The IOR held a meeting to celebrate its formation. The meeting was held as an online meeting broadcast using Go to Webinar. 101 members and guests gathered to hear Daniel Colbourne FInstR Arnold give his historical talk "The History of Flammable Refrigerants" The meeting was chaired by Dermot Cotter FInstR. The paper presentation was followed by an online discussion about the future of refrigerants. Questions and comments were sent in by Andy Pearson FInstR, Stephen Benton FInstR, Paul McGuckin, Neil Moss MInstR, Thore Oltersdorf, Malcolm Paterson MInstR, Eric Smith and Chris Rhodes MInstR. The meeting closed at 5.15pm.

3rd February 2022
The February meeting took place as a virtual meeting and was broadcast using Go to Webinar. The paper presentation "Improving Cooling and Oil Distribution in an Oil-Injected Screw Compressor" was given by Nausheen Basha winner of the Ted Perry Award. 31 people joined the webinar, and the broadcast was chaired by the Dermot Cotter FInstR, Chair of the IOR Papers and Publications Committee. Questions were asked by Mohammad Alshaikh Hasan, Lars Andersen, Dermot Cotter FInstR, Miriam Rodway MInstR and Mr Chaddha. The webinar ended at 4.57pm.

3rd March 2022
The March meeting took place as a virtual meeting and was broadcast using Go to Webinar. 86 members and guests join the live broadcast to hear the paper entitled "Ammonia Heat Pumps Helping to De-Carbonise Food and Beverage Industries"" given by Robert Unsworth MInstR. The meeting was chaired by the Dermot Cotter FInstR, Chair of the IOR Papers and Publications Committee. The presentation was followed by a discussion. Questions were submitted by William Thompson FInstR, Sean Hurley MInstR, Edward Clark, Barry Cooper MInstR and Adam Pawelas. The meeting concluded at 5.16pm

7th April 2022
The April presentation took place as a virtual meeting and was broadcast using Go to Webinar. 108 members and guests joined the webinar to hear the paper Carbon Dioxide Hazard Assessment and Safety System Requirement given by Dermott Cotter FInstR.
Questions and comments were submitted by Andy Campbell MInstR, Kamkari Babak, Bruce Nelson MInstR, Hanan Bareli MInstR, Luca Carettto, Marco Masini MInstR, Paul Cornwell MInstR. The meeting closed at 5.00pm.

The IOR undertakes a wide range of educational and scientific activities which provide benefit to the public and its members by promoting the advancement of the science and practice of refrigeration and air conditioning in all its applications for the good of society.

During 2020/21 the IOR adjusted its activities to minimise the risk presented by the global COVID-19 pandemic to its staff, members and its finances and to ensure the charity's continued long-term operation and ability to pursue its charitable objectives.

Mike Creamer FInstR

President and Chairman of the Board of Trustees

14th September 2021

The purpose of the IOR as set out in its Constitution is:

- The general advancement of refrigeration in all its applications, in relation both to the perfection of its methods, and to the extension of its services to the community
- To promote means for communication between members and their interchange of views
- To encourage invention and research in all matters relating to the science and practice of refrigeration
- To promote a sustainable approach to all aspects of refrigeration system design and operation
- To co-operate with educational institutions for the furtherance of education in the science and practice of refrigeration
- To hold meetings of the Institute for reading and discussing papers dealing with refrigeration and allied subjects
- To publish and distribute the proceedings or reports of the Institute
- To do all other things, incidental or conducive to the attainment of the above objects or any of them

Institute of Refrigeration Principal office
Kelvin House, 76 Mill Lane, Carshalton SM5 2JR UK

ANNUAL REVIEW 2020-21

1.1 Objectives and activities

The Trustees have reviewed these objectives and evaluated activities against them in line with the Charity Commission guidance on public benefit and confirm that the IOR continues to provide public benefit through:

- Activities to reduce the environmental impact of refrigeration, air conditioning and heat pump technologies by awareness raising of net zero objectives and providing high quality technical events and publications
- Promoting training standards, education, skills and careers opportunities by engaging with students, colleges, schools and the Institute for Apprenticeships and Technical Education
- Enhancing professionalism of individuals operating in the sector by offering ways of sharing information, encouraging Engineering Council professional registration, reviewing Apprenticeship Standards, recognising CPD (Continued Professional Development) of members, and promoting a Code of Professional Conduct.

The Trustees carried out a strategic review of activities necessary to achieve these aims during the year and confirmed the following priorities and targets for the next period which will have a long-term impact:

1. Leadership in education – supporting apprenticeships
2. Promoting our industry to schools & young people
3. Setting the environmental agenda for net zero
4. Supporting members – students, technicians and young people grades

The Impact and success in progressing these strategic priorities is reviewed at quarterly Trustee meetings where progress reports are presented, and activity adjusted accordingly. In addition, Trustees monitor Engineering Council professional registration levels, international partnerships with Institutes sharing our objectives, opportunities to host international research conferences in the UK, local and regional networking opportunities and providing public education and technical information on key issues such as safety, and environmental issues such as the use of low GWP alternative refrigerants and net zero cooling.

1.2 Signification activities and results

Technical publications

- Guidance Notes on CPD, STEM, Engineering Technician specifications
- Technical papers on heat pumps, carbon dioxide valve operation, integrated energy projects, low GWP refrigerants
- Updating of existing publications on refrigerant glide and cold store ceilings
- Translation of REAL Alternatives learning materials into additional EU languages
- Service Engineer Guides published or updated on risk assessment for A2L refrigerants, air conditioning technologies, optimisation of systems, brazing science, F-Gas bans.
- Conference Papers on Organic Rankine Cycles
- Annual Proceedings Journal

International profile

- Participation in an ASHRAE international Associations Alliance meetings
- Joint event with Institute of Refrigeration Ireland
- IOR was due to conclude the EU-funded element of a project managing a collaborative education project based on REAL Alternatives with a team of international stakeholders in June 2021. It will now support this as an independent international education activity.
- IOR participated in an international research project on the application of cryogenic energy storage in cold stores, responsible for communications. The project closed April 2021
- The IOR continued to support the UK membership of the International Institute of Refrigeration, an intergovernmental organisation
- A joint international conference was held in association with the international Knowledge Centre on Organic Rankine Cycle and International Institute of Refrigeration

Networking and exchange of information

- Free technical webinars were held on topics including heat pumps, community integrated energy, cold chain, elearning for alternative refrigerants, options for low GWP refrigerants, refrigerant monitoring, heat pumps and heat recovery, sustainable thermal systems, flammable refrigerant safety, cryogenics.
- The Women in RACHP network continued to promote diversity with career development webinars
- SIRACH webinars focused on key heating and cooling topics and research reports

- An international online conference was held on the topic of Organic Rankine Cycles
- Events were promoted from the IIR, BESA, Cold Chain Federation, Cryogenics Society, Institute of Refrigeration Ireland, ASHRAE, Refcold India, South West Society, CryoHub and REAL Alternatives.

- Regular online events allowed new members to gain an introduction to the benefits of membership, those interested in promoting the sector to find out more about STEM initiatives, how to apply for Fellow Grade, Engineering Council recognition, Mentoring Pathways for new members, reviewing the Apprenticeship Standard and discussion of Net Zero Refrigeration.
- Regular participation in ACRIB (Air Conditioning and Refrigeration Industry Board) meetings allowed IOR to contribute to the development of industry initiatives relating to education such as the ACRIB FGas Skillcard, to share technical information between industry associations and to monitor developments with legislation and policy in areas such as FGas Regulation, Hackitt review, Building Regulations and safety monitoring activity.

Education and Training

- Students on apprenticeship schemes were supported with free access to IOR publications and technical information, whilst some Training Centres were closed due to Covid-19 restrictions.
- Winners of the IOR RAC Student of the Year competition were interviewed and articles published in an IOR page in the RAC Magazine.
- The REAL Alternatives E-learning platform was finalised with online CPD for those who study the learning and several webinar events held to encourage take up of free elearning
- Activity to promote the take up of the IOR www.FantasticFridges.com website included webinars for STEM Ambassadors and the issuing of a summary of content for those giving talks to schools
- New technical guidance notes on Education were published on the topics of: CPD for members, becoming a STEM ambassador, giving talks to schools, Using Fantastic Fridges website, Engineering Technician Specification.
- A network of STEM Ambassadors and those interested in giving talks to schools was formed and webinars held to share experiences.
- A Steering Group of employers and trainers began a review of the existing RACHP Engineering Technician Standard, which underpins the Apprenticeship.

Encouraging and promoting innovation

- The J&E Hall International Gold Medal for practical contribution to the development of RACHP technology and Ted Perry Student Research Awards were not made during this period, as no IOR Annual Dinner was held which would have hosted the presentations, due to COVID-19 restrictions.
- The IOR continued to participate in the EU CryoHub project to promote the benefits of cryogenic energy storage for renewable refrigeration and power.
- The Beyond Refrigeration: Path to Net Zero initiative brought together expertise to support end users in their journey towards net zero and a group prepared a Template for Action.

The IOR benefits from significant contribution of volunteers, mainly members, through:

- Membership of Committees and Working groups to edit publications, peer review technical papers and organise conferences or events
- Technical talks at webinars
- Judging panels for awards and medals
- Reviewing and grading of membership applications

1.3 Achievements and performance

Significant activities generating income for the period included:

- Membership and Service Engineers Section fees and subscriptions
- An International Conference on Organic Rankine cycles
- Project Grants (CryoHub energy storage research project, SIRACH Sustainability Network, REAL Alternatives for LIFE)

Success indicators were monitored and reported on at Trustee meetings included:

- Membership levels
- Attendance levels at webinars
- Numbers of publications produced and downloaded
- Website and social media reach
- Engineering Council registrations

The IOR is a professional affiliate of the Engineering Council, which allows its members to register with the Engineering Council through a licencing agreement with CIBSE. The Engineering Council provides valuable benchmarks against which the Trustees evaluate IOR activity and standards in areas such as Codes of Conduct, CPD, appeals procedures, diversity, corporate status, membership profile and marketing.

1.4 Fundraising

IOR does not engage in public fundraising or grant giving. It incurs expenditure related to membership marketing and manages awards for innovation, some of which are funded by donation by external organisations. IOR members (individuals) pay annual membership fees.

1.5 Note on impact of COVID-19

The IOR made a rapid shift to providing online events and committee meetings during this period to ensure that services to members were not interrupted and to mitigate risk of COVID-19 pandemic to staff, volunteers, members and service users of in person meetings. Therefore, business disruption was minimised. This led to a reduction in costs (room hires, travel etc). There was a corresponding reduction in income and a small drop of 5% in membership due to external factors (i.e. business interruption or furlough of members). No changes in staffing levels resulted from this and no staff were furloughed. All Staff continued to carry out their duties working remotely from home during this period with all due security and GDPR processes in place. The office premises at Kelvin House were secured with alarm, CCTV and door cam monitoring as well as regular on-site checks by key staff.

1.6 Future planned expenditure

The small permanent secretariat at the IOR headcount was decreased during the year with the Project Officer contract concluding as EU projects came to an end. Further reductions in staffing costs are being pursued during the forthcoming year.

Key activities being undertaken in the next period requiring investment of additional resources are:

- International conference on the Cold Chain planned for Spring 2022
- Maintenance of REAL Alternatives elearning outputs and expansion to more countries
- Supporting strategic priorities including Beyond Refrigeration Net Zero publicity and technical output, encouragement, and training of STEM ambassadors to give talks in schools.

IOR.org.uk

REFRIGERATION AIR CONDITIONING HEAT PUMPS

The knowledge hub for refrigeration, air conditioning and heat pumps

The Journey to Net Zero

Dr Robert Lamb FInstR
Philipp Holzer MinstR
Graeme Fox FInstR
Prof. Neil J. Hewitt MinstR
Tom Southall

Why you should attend

- Find out what the key issues are around net zero for RACHP
- Hear from experts in different sectors what could be achieved in each area
- Explore options, tools and opportunities for you and your business to help deliver net zero to clients

Introduction

The drive to net zero is a global, national and sector challenge that is increasingly the focus of government policy initiatives and strategy for business and individual action. On 31 October – 12 November 2021 the UK will host the 26th UN Climate Change Conference of the Parties (COP26) in Glasgow which will bring parties together to accelerate action towards the goals of the UN Framework Convention on Climate Change.

This series of papers from the IOR looks at the options and opportunities for various sectors within the refrigeration, air conditioning and heating sector. It is part of the IOR's "Beyond Refrigeration – The Journey to Net Zero" initiative, that is developing:

- a template for steps to net zero designed to help users of heating and cooling evaluate where they are now and what they need to do in the short, medium and long term to progress on their journey
- resources such as case studies, tools, information about new technologies and techniques to help this journey – most of this produced by specialist organisations and institutions
- evolving policy papers to highlight to legislators incentives and regulatory areas that they could use to support positive change in the sector in areas such as balancing heating and cooling demands; and improving skills and attitudes
- a focus for activity around seven key cross sector areas of reducing the need for mechanical cooling, balancing heating and cooling demands, using energy intelligently, achieving best system performance, making the best use of available technology, developing the best people and skills and working together

This growing collection of material can be accessed at the IOR's micro website www.ior.org.uk/beyondrefrigeration

A major area of concern identified by the Institute in current national and international policy proposals, is the lack of recognition of the potential savings that could be achieved when both cooling and heating needs are considered together.

The Climate Action Pathway for Net Zero Cooling from the Carbon Trust lays out a vision for action on cooling co-authored by the Carbon Trust, Cool Coalition, Kigali Cooling Efficiency Program, Oxford University and High-Level Champions. It details the key milestones that will be necessary to fully implement the Paris Agreement, exceed the Kigali Amendment and contribute to the UN's Sustainable Development Goals in cooling terms. The accompanying Pathway to Net Zero Cooling Action Plan highlights in detail the areas where progress is needed and includes an action table.

The UK industrial decarbonisation strategy published in March 2021 does not specifically deal with RACHP, but covers key business sectors (such as metals and minerals, chemicals, food and drink, paper and pulp, ceramics, glass, oil refineries and manufacturing) that use substantial amounts of heating and cooling. UK Clean Growth and national Heat and Building Strategies are not expected to make the link between heating and cooling.

IOR members and their businesses have the expertise, practical experience, and professionalism to make a valuable contribution to the policy debate on how the UK can achieve low carbon heating and cooling. Uniquely, they also are responsible for providing the solutions to industry for current and future cooling needs – and for making sure that theoretical solutions work in practice and in the long term. Our sector is willing and ready to take an active role in addressing the issue of climate change, and this series of presentations and papers is helping to focus discussion and point the way forward for the RACHP sector.

The papers below provide insights on a sector-by-sector basis of our readiness to meet the UK's commitment to achieving Net Zero by 2050, identifying some of the key areas where progress is needed and which initiatives and tools are in place to help the industry achieve this important deadline:

- Industrial refrigeration
- Commercial refrigeration
- Air conditioning
- Heat pumps
- Transport refrigeration

Industrial Refrigeration

Dr Robert Lamb FInstR

The COVID-19 pandemic has highlighted the importance of industrial refrigeration in our modern society, delivering vital cooling for sectors including food production, temperature controlled storage, data processing and vaccine production. The projected growth of the world's population to close to 11 billion by 2100, alongside the associated demand for food, data, medicines and other products, suggests an increased requirement for cooling and emphasises the need to improve efficiency and reduce cooling density.

Progress to date

The environmental impact of industrial refrigeration is a combination of direct and indirect emissions. Over the past thirty years, the damaging effect of synthetic refrigerants including CFCs, HCFCs and more recently HFOs through direct emissions has led to global phase-out and phase-down strategies that remain ongoing. There has been a focus on reducing refrigerant leakage, the development of new fluids with lower global warming potentials and increased use of 'natural' fluids including ammonia and carbon dioxide (CO_2).

There have also been significant steps made in reducing indirect emissions through improved system efficiency. These have been achieved through a combination of product and system development, enhanced system control and the setting of efficiency standards. Product development includes improvements in compressor efficiency, enhancement in heat transfer performance and the use of energy reducing components such as electronic expansion valves. The use of variable speed drives for compressors, fans and pumps is now common place, delivering typical improvements in efficiency of 10% to 30% compared to fixed speed technology.

System development has included improvements to the transcritical CO_2 cycle, including the use of parallel compression and ejectors. Charge reduction has also been a focus, particularly through the development of low overfeed technology in ammonia systems and the cooling of secondary fluids via packaged chillers.

Controls systems have evolved to focus on delivering optimum efficiency throughout the year. Modern PLCs are able to integrate the benefits of variable speed technology and floating suction and discharge pressures to match changes in the cooling demand and ambient conditions.

The setting of minimum efficiency standards for chillers and refrigeration systems through the Eco-Design regulations is helping to drive the industry towards lower energy solutions.

What remains uncertain is what more is needed if we are to meet a net zero commitment.

The Future

Benchmarking

So where do we need to focus in the next three decades to meet the 2050 net zero target? A key first step is understanding what it will mean to be net zero. Each sector will need to define targets and consider how these will be measured. Benchmarking is likely to play an important role and an existing example involves the data collected for the temperature controlled storage sector (Evans et al., 2013), which includes surveys of more than 300 stores with data from a further 500 facilities. Figure 1 shows how energy consumption varies with store volume for chill, frozen and mixed temperature facilities. Recent data collected from a number of UK

Presented before the IOR on 4 November 2021

facilities indicates that the energy consumption of the majority of the cold stores were below the European average for a given volume and many are close to or below previous 'best practice' guidelines set by the Energy Technical Support Unit in 1994 (ETSU, 1994).

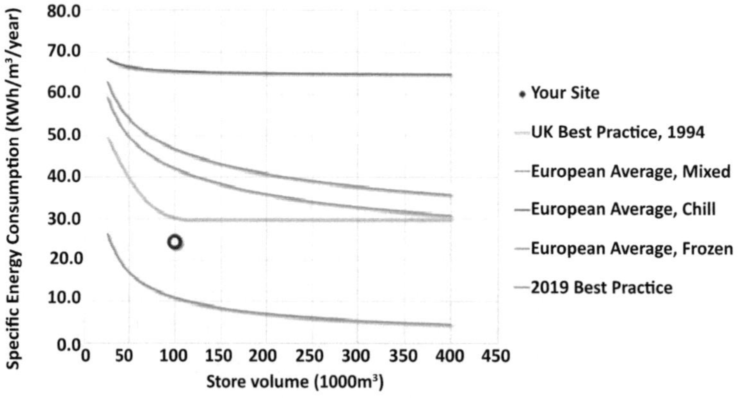

Figure 1: Specific energy consumption benchmarking

It is recognised that more needs to be done to achieve net zero and a new best practice line proposed from data taken for a number of modern cold and chilled storage facilities (Pearson 2019) suggests a significant reduction in energy consumption of up to 30% of the accepted ETSU figures is possible through better building and refrigeration plant design. This step change in expectations is an example of what is needed as we move to net zero. Whether this new best practice benchmark will be enough needs to be investigated and it could well be that further improvements are necessary. It will be important to look at similar benchmarking for other sectors and this may be driven by legislation as the government in England and Wales work towards setting minimum energy performance standards for industrial buildings based on actual annual consumption.

Performance analysis

Seasonal efficiency performance standards are already in place under the Ecodesign regulations and should continue as most industrial refrigeration systems work at part load capacity and in lower ambient temperatures for the majority of the year. It is important that this isn't just a paper exercise at the point of selecting equipment and the focus must be on monitoring performance and making any necessary adjustments to optimise efficiency over a system's operational life. Cloud based technology makes it possible to monitor performance data in real time, compare the findings with the theoretical design performance and identify where there are gaps. It is also possible to highlight what needs to be done to 'plug the gap' and address a shortfall in performance maintenance, repair or changes in control. By doing so, energy consumption can be optimised all year round. Using this type of data analysis also enables equipment performance and reliability to be monitored to ensure maximum longevity.

Heat recovery

The majority of energy used during cooling processes is discarded to the environment through condensers or cooling towers. By capturing and boosting this waste heat, industrial refrigeration systems can play a positive role in decarbonising our future heating demands and avoid the need to burn gas. Implementation of heat recovery and heat pump systems similar to that in Figure 2 will enable waste energy to be captured and used for hot water heating and other heating processes. Applications generating water at +90°C are in operation today and development of higher temperature systems is ongoing. For sectors where there are large heating and cooling demands such as food production, energy recovered from refrigeration plants and boosted by electrically driven heat pumps can be used to heat water for CIP and other process requirements. For businesses with a large cooling demand but a small heating demand (e.g. temperature controlled warehouses, data centres) waste energy could be recycled into district heating networks for use by others and by doing so, reduce emissions elsewhere.

Figure 2: Packaged industrial heat pump in manufacture

Conclusion

Positive steps have been taken to lower the environmental impact of industrial refrigeration by reducing direct and indirect emissions to the atmosphere but more will be needed as a growing population will increase demand. The focus will be on a combination of improved equipment/system design, better analysis of operating data and enhanced recovery of waste heat to support the decarbonising of heating.

References

Evans, J.A., Huet, J-M., Reinholdt, L., Fikiin, K., Zilio, C. Houska, M. Landfeld, A. Bond, C. Scheurs, M. van Sambeeck, T.W.M., 2013. *Cold Store Energy Performance*, Proceedings of the 2nd IIR Conference on the Cold Chain and Sustainability, Paris, IIF/IIR, 310-317

ETSU, 1994, *Energy Consumption Guide 37: Cold Storage Section*. Energy Efficiency Office, Department for the Environment, Harwell, United Kingdom

Pearson, A., 2019, *Energy Performance of Industrial Cold Storage Facilities*, Proceeding of the 25th IIR International Congress of Refrigeration , Montreal, Canada

Achieving Net Zero by 2050 – Commercial Refrigeration

Phil Holzer

Abstract

This paper discusses the current initiatives and the future challenges in achieving net zero in retail commercial refrigeration systems by 2050. It focuses on the decarbonisation of heating within stores, the switch to heat recovery and use of heat pumps. The paper outlines a number of the technologies actively being developed by retailers to reduce energy demand and concludes with an assessment on the likelihood of achieving the 2050 target.

Keyword: Commercial refrigeration, Net Zero, retail

Introduction

"The climate emergency is recognised as one of the greatest threats to our planet. The UK government is participating in global efforts to limit the global temperature rise to 1.5°C and has set legislative targets for the UK to reach net zero greenhouse gas emissions by 2050. UK retailers recognise and support the need for urgent action" [1]. As part of the decarbonisation of heating systems and the switch to electric heating, retailers are looking to recover waste heat and generate additional top-up from refrigeration systems utilising ultra-low GWP refrigerants. Systems must holistically consider the operation of both heating and refrigeration to ensure efficient operation and prolong the service life of equipment.

Background

Retailers are generally targeting net zero emissions in their own operation well in advance of the 2050 deadline [2]. The last decade has already seen a substantial reduction (circa 54%) in the carbon footprint of most retailers [Figure 1]. The largest part of the reduction is as a result of the switch to renewable energy and the replacement of HFC with CO_2 refrigeration systems.

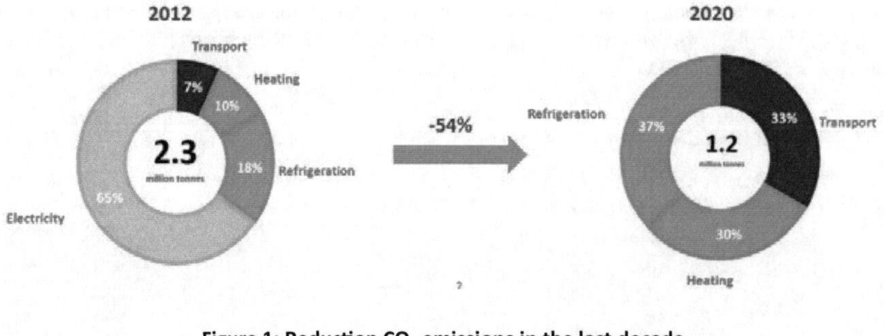

Figure 1: Reduction CO_2 emissions in the last decade

All retailers are committed to sourcing 100% of their electricity from renewable sources by at least 2035 [3]. Renewable energy is sourced in three ways. Either on-site, off-site generation or from certificate backed suppliers (REGO) [4]. Retailers are looking to increase both on and off-site generation [5] and reduce reliance on REGO backed sources. Therefore, the indirect emissions have largely been removed and the focus is now around the reduction of direct emissions. Refrigeration systems account for some 37% of this.

Presented before the IOR on 4 November 2021

Switch to low GWP refrigerants

Replacement of HFC systems with transcritical CO_2 (GWP of just 1) is now commonplace and well established, having been undertaken for the last 5 years. The capital cost is more or less equivalent to HFC systems, due to their increased saturation in the market. Service engineers working in the retail sector have been upskilled and are now widely experienced in CO_2 systems.

Typically, a quarter of the retailer's estate have CO_2 refrigeration systems. There are significant challenges in the replacement of the remaining HFC systems ahead of 2035. Some of these include the availability of labour, material, and manufacturing resources and of course the capital budgets to deliver on the 2035 goal.

Decarbonisation

Retailers are also planning the decarbonisation of heating systems by 2035. Effectively requiring the replacement of all gas fired boilers and a switch to electric heating. This presents further challenges, most notably around the electrical supply infra-structure, as the demand for EV charging rises. To enable this expansion, it is critical that we consider the following:

- The capture of waste heat from refrigeration systems and its integration into the Heating and Ventilation plant. CO_2 systems provide a great opportunity to not only recover waste heat, but also to generate increased heat output during periods of high heating demand.
- The optimisation of all parts of the refrigeration and heating system to reduce energy consumption.

Heat recovery

Heat recovery from refrigeration systems is crucial, although its application should be carefully considered. Artificially loading refrigeration systems or the use of false load systems may not be the most efficient way of generating this heat in all conditions.

Refer to Appendix A. It provides an example of the heat recovery mode vs store heating demand and calculated cooling COP. Heating demand can be met by the refrigeration system giving up its free heat, running down to 17°C ambient. This maximises the system efficiency.

Table 1 indicates some of the considerations.

Table 1: Heat recovery modes of operation

Heat recovery mode of operation	Pros	Cons
Artificially operate the systems transcritically in low ambient temperatures, to generate increased heat output.	Reduces the amount of top-up heat required from other sources and reduced capital cost.	The COP will dramatically reduce. Typically consuming nearly three times as much energy, when operating sub-critically. Service life of the refrigeration system is likely to be reduced. Maintenance costs will be higher.
Extracting free heat from the refrigeration systems without elevating discharge pressures.	Higher system COP can be maintained. Increased service life Lower Opex cost	Limited heat output at very low ambient temperatures but enables higher COPs to be maintained. Increased capital spend as additional heat pump capacity required

Heat recovery plates need to be carefully selected to limit pressure drop near peak design conditions but selected at an optimal point to match the heating demand of the store. There is little point in providing a plate that can generate a heat output beyond the store's heat demand. The design of the heat recovery plate also needs to consider thermal stress, typically more than 60k between the gas inlet temperature and water outlet. This can readily be exceeded when running transcritically.

Heat recovery from refrigeration systems are unlikely to meet the store's heating demand on their own, therefore additional electric heating or heat pump systems will be required for top-up.

Although heat pumps may provide improved efficiency at certain conditions over a transcritical CO_2 system, the current technology has them operating on R32 with a GWP of 675 or HFO refrigerants with high embedded GWP. Even such relatively low GWPs are unlikely to help with the journey to NetZero and likely to be legislated against. CO_2 heat pumps are being developed by the major heat pump manufacturers. In the interim, an alternative source for CO_2 heat pumps may be the current refrigeration plant manufacturers, although they are unlikely to be able to compete with the high-volume heat pump manufacturers once they reach the market.

The integration of the Store's BMS controls with the refrigeration system can be difficult to facilitate efficient heat recovery. The BMS system normally signals the store's heating demand but does not really consider the most efficient way to operate the refrigeration system. Equally the "off-the-shelf" refrigeration controls can be very crude in their application of heat recovery, offering little flexibility to optimise the operation of the entire system. In addition, we must have control systems with open protocols to allow data to be drawn from multiple systems and allow inter-communication.

Future heating, ventilation and refrigeration systems will be integrated into packed units to provide a much more efficient heat and cooling system. Such systems are beginning to become available in the market. Such combined systems will require service technicians to be crossed trained in both the heating and refrigeration systems. SLAs and call-out responses will need to be aligned.

The switch from gas heating to heat recovery has some additional challenges around scheduling. HVAC and refrigeration plant replacement have rarely been aligned. Retailers are now fitting heat recovery to new refrigeration systems as standard, making stores effectively ready for a future gas boiler replacement and heat recovery utilisation.

Other areas for improved refrigeration efficiency

Most of the low hanging fruit around the use of EC fans, LED lighting and application of off-cycle defrost have already been taken. There are a number of other areas for improved energy efficiency. They are beyond the scope of this paper to detail, but worthy of further development and consideration.

- The use of doors on dairy and meat cabinets is widely used in small format stores, generally up to a maximum of 20,000 ft^2 of net sales area. The retailer's merchandising team normally object to their use in larger stores, citing the impact on the customer experience and sales. Refrigeration load is typically reduced by some 30-40%. Future limits in the electrical infrastructure may force doors to be more widely adopted. Their use would also negate the requirement for cold air retrieval (CAR) and heating offset in lower ambient temperatures. This can represent as much as 25-30% of the store's heating demand.
- The use of stepper EEVs over pulsing type can provide a more stable suction pressure and resultant energy benefit. This is well established on HFC systems, but not yet proven on CO_2.
- The use of cabinet "shelf edge technology" is now commonplace, providing reduced energy consumption of circa 5-7%. It also reduces air spillage and heating demand.
- A number of cabinet manufacturers are developing cabinets which will substantially reduce the connected load (up to 50%) with increased evaporating temperatures (up to 0°C).

These examples will of course reduce the potential heat output from the refrigeration systems, but their energy benefits can not be ignored, particularly when they also reduce the CAR heat demand.

Retailers should no longer be hung-up on a single controller manufacturer, but instead define the requirements and design the network infra structure to allow a simple switch to the latest most efficient technology.

Conclusion

The journey to NetZero is very much at the forefront of the retailer's core purpose. They are generally engaged with the latest technological advances and willing to assess and validate their performance in real world terms.

Budgets for the replacement of HFC systems were established a number of years ago and will enable their phase-out by 2035. Additional funding to support the decarbonisation of heating is also planned and fixed but closely wed with the installation of new CO_2 systems.

The traditional heat pump manufacturers must bring natural, sub-150 GWP refrigerants to the market within the next 2-3 years.

Controls manufacturers must provide open communications protocol devices that cross both heating and cooling markets to ensure heat is generated in the most energy efficient methods. Retailers have a responsibility to encourage, invest and challenge manufacturers. They should not accept standard offerings.

Retailers are generally in a good position to deliver Net Zero ahead of 2050.

[1] https://brc.org.uk/media/677373/climate-action-roadmap-executive-summary-apr21.pdf

[2] https://www.about.sainsburys.co.uk/news/latest-news/2020/28-01-20-net-zero

Presented before the IOR on 4 November 2021

[3] Tesco is committed to sourcing 100% of its electricity from renewable sources by 2030 – to include over 50% from power purchase agreements and on-site generation. (https://www.there100.org/re100-members)

[4] https://www.ofgem.gov.uk/environmental-and-social-schemes/renewable-energy-guarantees-origin-rego

[5] https://www.edie.net/news/10/Tesco-to-support-creation-of-three-UK-solar-farms/

Appendix A: Example of heat recovery mode vs store heating demand vs system COP

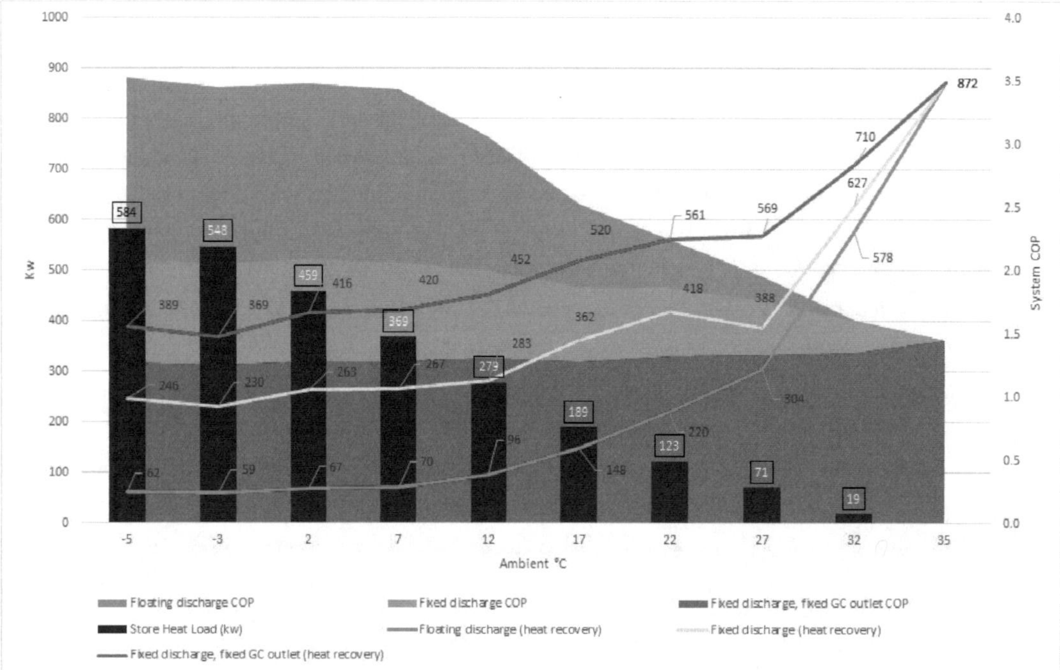

Air Conditioning

Graeme Fox FInstR

The term "air conditioning" is often misleading, poorly applied, and is frequently used to discuss comfort cooling when, in fact, the term covers far more than simply direct expansion cooling products.

Some would define it as "the process by which we create and maintain certain temperatures, relative humidity and air quality in indoor spaces", some that it is "the process of removing heat and controlling the humidity of the air in indoor spaces to create a more comfortable environment". Neither of these is wrong, but similarly neither is correct either.

These distinctions and definitions become extremely important, however, when we are looking at air conditioning in terms of our Net Zero ambitions and where the technology aligns with those ambitions and the road map we need to follow to achieve them. For example, when does an air conditioning system become accepted as a heat pump? Or conversely, when is a heat pump an air conditioning unit?

Aware that heat pumps are being covered in a separate session I am keen not to create too much crossover here, but it is important to not focus too much on one aspect of "air conditioning" to fit an agenda simply because we are the Institute of Refrigeration. In terms of Net Zero achievement the inclusion of air conditioning systems as valid heat pumps which may include a cooling functionality, or are comfort cooling systems with a reverse cycle functionality, is an absolutely crucial distinction.

Government officials and lobbying groups appear to be largely aligned now in the acceptance that upwards of 600,000 heat pumps need to be installed annually, with that figure growing year on year, if we are to meet our Net Zero by 2050 ambitions. But that will not be achieved by only considering air to water heat pumps for the domestic sector. The Insulate Britain demonstrations on the M25 and beyond earlier this year, disruptive to business and society as they are, are not without good intentions. Their concept is that domestic buildings are often so poorly insulated that the application of air to water heat pumps often results in poor performance and high energy bills for the home owner or tenant – and the net result of that is simply a bad reputation for the technology itself. Yet they fail to recognise that over insulating these homes creates poorly ventilated buildings that overheat in warmer months – particularly as the effects of climate change increase – and they create situations where the indoor air quality can become toxic to the occupants.

So if the increased insulation of homes is creating a situation where those homes are overheating and the air quality is becoming poor, what can be done about that? The answer of course is that air conditioning is needed to ensure that the quality of the air inside buildings is fit for habitation or occupation.

At rest, the average adult inhales and exhales around 13kg of air per day, compared to around 2kg of food and 3kgs of fluids – yet health advice surrounds us on the subject of food and drink and very little on the subject of air quality. In the UK we spend an average 90% of our time indoors, so the quality of the air we breathe is hugely important in terms of long term health and wellbeing of the nation as a whole.

Controlling CO_2 levels indoors as occupancy numbers increase is a vital way of demonstrating at the most basic level that sufficient air changes are in place; but this cannot mean simply installing a ventilation system which supplies fresh air and/or expels foul air. With an eye on our nations's energy consumption – and of minimising energy use to help achieve Net Zero emissions – we must be looking at mechanical ventilation with heat recovery more frequently as a sustainable solution.

Looking then at these systems and how they work, we see the heat recovery aspect is achieved by recovering sensible heat out of the return air, transferring it through heat exchanger filter media, and re-applying that sensible heat to the fresh or outside air to ensure the air being drawn into the building is not much lower than that being removed. 100% fresh or outside air, but treated to maintain internal temperature and humidity levels and filter out particulate matter from the make up air. That air being supplied is therefore conditioned air, by almost any definition you would use for air conditioning, and yet no refrigeration fluid or process has taken place.

Additional levels of filtration can also ensure that particulate matter (PM) created by cooking or from fibres shedding off furniture and clothing is kept low, that volatile organic compounds (VOCs) aerated by cleaning products or from printers are removed from the air that we breathe, and that CO_2 levels are maintained at healthy levels well below 1200 ppm.

Our Government's Net Zero ambitions will see an enormous impact on our lives at work and at home in the coming years – energy bills will continue to rise as the energy suppliers seek to recover infrastructure costs of implementing an increased demand on the electrical grid needed to replace the increasingly out of fashion gas network. The climate will continue to change regardless of these actions and for the foreseeable future that change will be in the form of a general warming of the Earth's atmosphere. Homes and workplaces overheating will become an ever more critical problem which will demand some form of mechanical comfort cooling for short periods in buildings where the natural ventilation route isn't an option.

So the application of air conditioning systems in the broad sense of the term is crucial to our targets being met, and the availability of a cooling function for short periods of the year where needed should not exclude the reverse cycle heat pump air conditioning units from being considered as a major piece of the Net Zero puzzle.

Heat Pumps

Professor Neil J. Hewitt, MInstR

Introduction

The electrification of space and domestic hot water heating is the most efficient use of our dominant renewable energy source i.e. wind. Heat pumps provide high efficiency, can be retrofitted directly into homes and are up to six times more efficient than the substitution of natural gas with green hydrogen. However, the electrification of heating has revealed challenges with heat pumps, i.e. their end-user acceptability, capital cost, operating cost, scalability, electricity network capacities, heat pump efficiencies and potentially higher temperature requirements for existing buildings. Thus, there are geographical factors that limit the just transition to heat/grid co-decarbonisation through addressing (i) inadequate building thermal quality; (ii) inadequate electrical network capacity to address electrification of heating; (iii) insufficient understanding of stakeholders and their needs to facilitate implementation; and (iv) insufficient understanding of the interaction between these three aspects. Therefore, improved Electric Heat Pump (EHP) technologies that address heat and hot water demands for buildings will be presented alongside the potential of EHP/thermal storage for buildings.

Refrigerants

Given the expected downward trend in single family home heat distribution temperatures (through hoped for insulation improvements and new builds), the heat distribution temperatures are expected to drop for such systems from over 60°C to 40°C or less. Regarding refrigerants, low global warming potential (GWP) is an important factor. EU F-gas regulations will reduce the current dominant hydrofluorocarbons (HFCs) and will challenge R410a, the prevalent residential heat pump working fluid. Reduced GWP R452B and R454B along with R454B show marginally superior coefficients of performance (COP) [1] while R32 is becoming popular. However, some R410A replacements with a lower GWP (<150) report COPs reduced by 4% to 14% so drop in alternatives with low GWP may not be feasible [2]. Natural working fluids e.g. R290 (propane) have similar performance in purpose designed systems [3] and there are several significant manufacturers. R744 (CO_2) is growing in popularity for water heating/air heating systems via a transcritical cycle but with only a small number of residential scale manufacturers. Regardless of the choice of fluid, in changing from R410a, it appears that a low GWP working fluid will be challenging to find. R466A (R32, R125 CF3I, or trifluoroiodomethane) with a GWP of 733 appears one of the most promising in terms of performance with a 5% increase in performance [4]. Therefore, specialist fluids (R744) requiring a significant system design shift or lower density refrigerants must be utilised e.g. R290 (flammability), R444B (lower flammable, GWP = 295) or R455A (lower flammable, GWP = 146). As previously stated, R290 is well known but the relative performances of the other lower GWP alternatives are less so. R444B shows a 4-5% increase in performance over R407C [5]. R455A will have a lower performance that R410A. So R290 may become the fluid of choice alongside a limited number of milder flammability low GWP mixtures as current refrigerant chemistry has been largely explored.

Heat pump components

Considering the main mechanical components of the heat pump, system noise through key legislation comes from the UK BS4142 for example. Regarding BS4142, 35 dB(A) to 40 dB(A) at night is "typical" suburban noise level. There is also a need to add 5 dB(A) for a "tonal" quality of fan and if the sound source is 10 dB(A) above background noise "complaints may be expected". IEA Heat Pumping Technologies Annex 51 "Acoustic Signatures

Presented before the IOR on 4 November 2021

of Heat Pumps" sees over 60dB(A) for operation and defrost activities [6]. Therefore, selection of acoustic covers, fan types and operating speeds, compressor speeds, pipe sizing (velocity) etc. must be considered. However, variable speed drives (compressors, fans, pumps) are necessary components in energy efficiency and in demand side response roles (when accompanied by household heat distribution controls). The concept is simple in that the thermal inertia of the building can provide a portion of demand side response (through a heat pump "off" or "throttling" phase) with throttling of the heat pump leading to 40% to 65% load reduction depending on the building and outside air temperature [7].

Thermal storage optimisation

Thermal storage sizing will be dependent on several factors including integration of household variable non-dispatchable renewable energy (e.g. solar), weather/climate, integration of local electricity network variable non-dispatchable renewable energy (e.g. wind), the impacts of a weak electricity network in terms of the electrification of space heating and the electrification of transport. In addition, for thermal storage, there are challenges of compactness in retrofitting air-source heat pumps and thermal stores in existing single-family homes and potential acceptance of aggregation control of operations in terms of electricity network capacity management. Regarding building integrated renewable energy e.g. solar, the challenges are to potentially increase self-consumption in terms of electricity (Photovoltaics) for personal economic purposes and/or limiting export to the local electricity network. The latter has ramifications for local electricity network capacity and for electricity voltage transformer sub-station abilities to reverse flow. Heinz and Rieberer [8] simulated a retrofitted single-family home in Zurich, Switzerland with PV, an air-source heat pump, and thermal storage options (water) ranging from 500 litres to 2000 litres. The proposed retrofit reduced the heating demand by 27% and concluded that "existing" radiators were a better mechanism for addressing PV integration. Thus, 5 kWp PV and a 1 m^3 thermal energy store was the most economic for a 185m^2 single family home. Insulation is important and its inclusion reduced the needs for buffer storage while battery inclusion serves the role of local building self-consumption of PV generated electricity. Shah et al [9] successfully operated a higher temperature cascade air-source heat pump and 600L of water based thermal storage responding to wind availability in a detached family home without fabric retrofit and utilising the existing hot water radiator system originally designed for a gas boiler with flow and return temperatures of 80°C and 60°C respectively.

Discussion and Conclusions

When considering the electrification of heating, actions that mitigate negative impacts on the electricity network must be prioritised. Therefore, reduction of thermal demand (e.g. building insulation) will reduce air-source heat pump electricity demand and it has been shown that it also reduces thermal storage size [9]. The electrification of heating in areas of weak electricity network will need to consider diversity factors, aggregation and potentially their role electricity network ancillary services. However, demand forecasting and status of individual heat pump and thermal energy storage systems to meet that demand is not well established, as are the relevant communications and demonstration of revenues e.g. against infrastructural upgrade deferral for example.

References

[1] *Drop-in Performance of the Low-GWP Alternative Refrigerants R452B and R454B in an R410A Liquid-to-Water Heat Pump*, Applied Thermal Engineering16 September 2020, Jaime Sieres, Ignacio Ortega, Estrella Álvarez

[2] *Evaluation of Low-GWP and Mildly Flammable Mixtures as New Alternatives for R410A in Air-Conditioning and Heat Pump System*, International Journal of Refrigeration 28 September 2020, Binbin Yu, Hongsheng Ouyang, Jiangping Chen

[3] *Vapour Compression Heat Pumps with Pure Low-GWP Refrigerants*, Renewable and Sustainable Energy Reviews 28 November 2020, Di Wu, Bin Hu, R. Z. Wang

[4] *Energetic Performance Analysis of R466A as an Alternative to R410A in VRF Systems*, Engineering Science and Technology, an International Journal, Volume 23, Issue 6, December 2020, Pages 1425-1433, Atilla G. Devecioğlu, Vedat Oruç

[5] *Low GWP R22 Replacement for Air Conditioning in High Ambient Conditions*; International Journal of Refrigeration, September 2015, Ankit Seth i, Elizabet Vera Becerra, Mark W. Spatz

[6] https://heatpumpingtechnologies.org/annex51/wp-content/uploads/sites/59/2020/11/iea-hpt-annex-51-d22-v1.pdf

[7] *Large-Scale Demonstration of Precise Demand Response Provided by Residential Heat Pumps*, Applied Energy 8 February 2019, F. L. Müller, B. Jansen

[8] *Energetic and Economic Analysis of a PV-Assisted Air-to-Water Heat Pump System for Renovated Residential Buildings with High-Temperature Heat Emission System*, Applied Energy, 27 April 2021, Andreas Heinz, René Rieberer

[9] *Energy and Economic Optimization of Solar-Assisted Heat Pump Systems with Storage Technologies for Heating and Cooling in Residential Buildings*, Renewable Energy 1 May 2020. Maria Pinamonti, Paolo Baggio

The Journey Towards Emission Free Temperature-Controlled Distribution On Road Vehicles: An Industry Plan

Tom Southall

Introduction

The UK Government's ambition to deliver a net zero economy by 2050, as well as the interim target of reducing emissions by 78% by 2035 will have profound impacts on all industrial processes. To meet this goal, industries are expected to undergo rapid change. As well as reducing Greenhouse Gas (GHG) emissions such as CO_2, there is an urgent need to reduce the concentrations of pollutants known to cause detriment to human health emitted from our vehicles such as NOx and Particulate Matter (PM), particularly in crowded urban centres.

Temperature-controlled distribution is critical to the UK supply chain, ensuring perishable products including food and pharmaceuticals are transported safely across the cold chain and reach the consumer in the safest condition. It exists at every stage, from the farm to our homes, hospitals, restaurants and retail spaces. The cold chain is also a critical tool in national decarbonisation efforts – driving down postharvest food waste and the associated emissions of over production by preserving food quality for longer.

Despite these benefits, temperature-controlled distribution is also responsible for a range of GHG and air quality emissions emitted not only from a vehicle's primary engine, but also uniquely to the cold chain, from the operation of refrigeration units on vans and Heavy Goods Vehicles (HGVs) known as Transport Refrigeration Units (TRUs).

Emissions from temperature-controlled distribution come from four main areas (as detailed in Figure 1). Whilst all sources are critical to the overall decarbonisation of road haulage, this new report focuses on the specific emissions added by the operation of refrigeration units on vehicles.

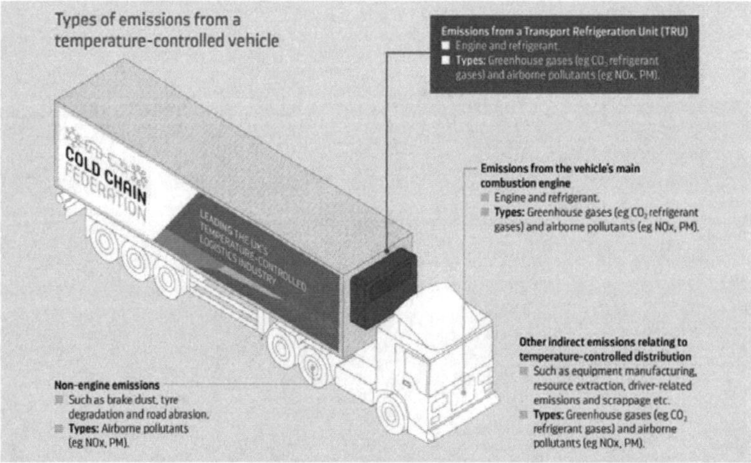

Figure 1: Types of emission from a temperature-controlled vehicle

The Cold Chain Federation and its members are committed to the goal of removing the reliance on diesel in transport refrigeration, however there are significant challenges which need to be overcome. The Journey Towards Emission Free Refrigeration on Road Vehicles is our vision for how a collaborative partnership between

Presented before the IOR on 4 November 2021

industry, Government and innovators, supported by robust policy and economic support can achieve emission-free transport refrigeration in line with the UK's net zero by 2050 ambition.

A brief summary of the report is presented below. The full version can be downloaded at: https://www.coldchainfederation.org.uk/cold-chain-net-zero-project/

The outlook for eliminating emissions from TRUs

There is a clear pathway towards refrigeration powered from the vehicle's main engine for fixed LGVs and rigids, but for these vehicles, completely removing emissions will be reliant of electrifying the main engine. 'Hybrid' systems are also increasingly being used for articulated HGVs to allow for the trailer's fridge to be powered from the HGV engine. This solution has benefits for air quality emissions but is still reliant on the diesel engine of the HGV. Other options exist for reducing emissions from the current fleet of TRUs such as HVO fuel.

The plan sets out that prior to the successful decarbonisation of HGV trucks, potentially quicker emission reductions will be achieved by powering the TRU from emission free sources, such as electric – in the form of 'smart batteries' or potentially hydrogen and nitrogen as this technology is developed. Electrifying the trailer has the additional benefit of being conducive to the current supply chain, with trailers charging at depots whilst they are being loaded/unloaded.

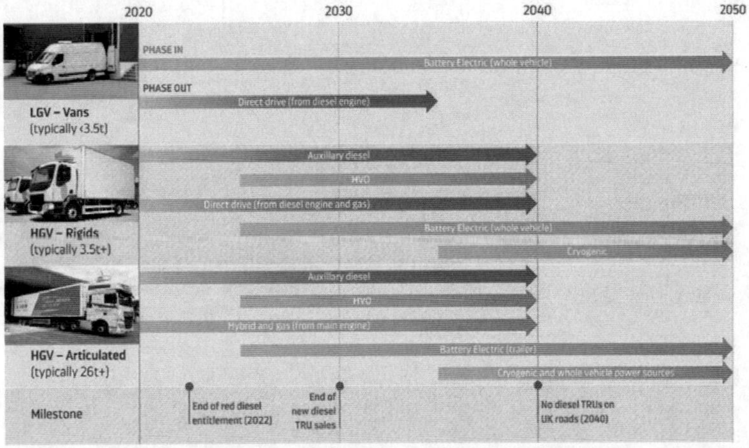

Figure 2: Technological outline for TRUs

Prior to 2030 the industry must look to reduce emissions from existing technology whilst trialling and building confidence in developing technology to become emission free by 2040.

Short term: by 2030 (transitioning to lower or emission free refrigeration)	Long term: by 2040 (move to emission free refrigeration)
Vans and small rigids	**Vans and small rigids**
■ As more vans upgrade to at least Euro VI (or later) standards, those operating 'direct drive' systems will reduce emissions significantly. ■ Increasing adoption of electric whole-vehicle solutions to achieve a fully electric refrigerated vehicle. ■ Reduction in the GWP of gases used as a refrigerant due to F Gas regulations.	■ All vehicles will run their refrigeration systems from the main engine of the vehicle, which is likely to be electric by 2040. ■ Emission free, or ultra-low refrigerants replacing high GWP gases.
Large rigids and articulated	**Large rigids and articulated**
■ Improved efficiency in operation of refrigerated trailers to significantly reduce the overall emissions from auxiliary diesel TRUs. ■ Some adoption of lower emitting fuels to reduce emissions, such as HVO and biofuels. ■ Increasing adoption of hybrid technology to run systems from Euro VI (or later) engines where possible. ■ Some adoption of 'smart battery' powered trailers as this technology becomes more viable. ■ Reduction in the GWP of gases used as a refrigerant due to F Gas Regulations.	■ No auxiliary diesel TRUs for large rigids, which will be able to be electrically powered by the main engine when in motion and by plug-in electric when stationary (or from batteries). ■ Widespread adoption of smart battery technology to achieve mostly diesel free emissions from refrigerated trailers. ■ Alternative technology such as nitrogen fully scoped and beginning to be adopted if viable. ■ Emission free, or ultra-low refrigerants replacing high GWP gases. ■ Decarbonisation of HGV engines could result in wider adoption of a single vehicle (truck and trailer) power source for both motion and refrigeration.

Figure 3: Outlook for eliminating emissions from TRUs

What can be achieved?

Based on the findings of the report, we present four industry targets for removing emissions from TRUs:

INDUSTRY TARGET 1	INDUSTRY TARGET 2
No transport regrigeration units (TRUs) to be sold into the UK market containing refrigerants with a GWP of more than 300 by 2025.	No transport refrigeration units (TRUs) should use refrigerants with a GWP of more than 300 by 2035 (in line with the Kigali agreement).
INDUSTRY TARGET 3	**INDUSTRY TARGET 4**
Ambition that no new diesel TRU should be placed on the UK market after 31 December 2029.	No vehicle operating on UK roads should be using a diesel powered TRU after 31 December 2039.

Figure 4: Industry targets for removing emissions from TRUs

Barriers to delivery of these targets:

- Infrastructure – provision of required charging or fuelling infrastructure for electric, or hydrogen/nitrogen fuels
- Maintenance, reliability and operator confidence – including training engineers for alternative technologies
- Affordability and the case for investment – the significant cost differential of developing technology and supporting infrastructure relative to traditional diesel TRUs

How this plan can be achieved – industry led action, supported by effective Government regulation and investment

Operators should:

- Continue to reduce air quality and GHG emissions from TRUs through more efficient usage of existing equipment whilst trialling and adopting lower emitting, or emission-free alternatives to diesel and high GWP refrigerants as they come to market and are financially viable.

TRU manufacturers should:

- Continue to reduce the use of high GWP refrigerants in line with F Gas Regulations and seek to increase the pace of innovation in design aimed at reducing air quality and GHG emissions.

Trailer and refrigerated van manufacturers should:

- Continue to push the limit of trailer design to maximise thermal efficiency and continue to seek innovative solutions to generate emission free power from the trailer's roof, axles and brakes.

The Cold Chain Federation will:

- Consult on the introduction of industry-led scheme to promote best practice in efficient temperature-controlled distribution.
- Produce a best practice guide for businesses and operators to maximise the efficient operation of TRUs and reducing associated emissions.
- Continue to support efforts to quantify emissions from temperature-controlled distribution (as detailed in our 'Defining a Net Zero Cold Chain' document) and to track progress of this plan through regular reviews. The first review will be completed no later than December 2025

The Government must:

- Provide confidence and a clear deployment strategy for the technologies which will be backed to become widespread in the future to give operators confidence in major investments in developing technology e.g. electric, hydrogen, biofuels.
- Provide a clear investment strategy to facilitate the improvement of electric charging for vehicles and temperature-controlled trailers at depots and at rest points along the road network which must include subsidies or grants to overcome the significant costs for operators of installation of equipment.
- Support the targets set out by this plan with effective regulation and policy to achieve successful implementation (see report for full details of policy recommendations)
- Support the investment in trials, or adoption of lower and emission free technologies through Green Funds, extension of super tax credits, Grants or other tax allowances and ensure that research funding is allocated to realistic solutions.
- Ensure the requirements of temperature-controlled operators are taken into account when developing the freight system of the future as outlined in the Transport Decarbonisation Plan.

The full report, The Journey to Emission Free Refrigeration on Road Vehicles can be downloaded from the Cold Chain Federation website, along with other publications and information on our Net Zero Project: https://www.coldchainfederation.org.uk/cold-chain-net-zero-project/

DISCUSSION REPORT

Graeme Maidment, chairing, thanked the authors and invited questions.

Nick Franzen
Excellent presentation. A lot of work and a lot of studies have gone into investigating plant efficiency. My question is about effect efficiency and evaporators. The evaporator tends to have a double impact i.e., you use energy to form ice and then you use more energy to defrost it, same with evaporator fans. If you oversize the fans you use more energy running the fans and you fail to extract the energy from the cold store. I wondered what your thoughts were on load efficiency?

Rob Lamb
One of the things that we have been looking at in terms of plant efficiency is finding the most efficient way to size and operate evaporators fans. Historically fans are fixed speed and therefore typically oversized. The move to EC fans offers a big benefit because you can actually set the speed to what the store needs. Taking this a step further, the fan speed can be varied based on cooling demand in the building and this delivers a significant benefit. It considers temperatures throughout the building and then varies the fan speeds for individual areas to meet the cooling requirement. For example, if the temperature all the way to the doors is okay, you can reduce your fan speed and save energy. That's the bit of the integration I was mentioning within the presentation earlier. It's actually the control system recognising that there is a better way to operate the evaporator and at the same time looking at compressor control and identifying that, if I do this here, this is the effect elsewhere. Condensers are another example. There is a point with chillers where it is beneficial to change the fan speed, to run the fans at high speed and save energy on the compressor. In lower ambient conditions and at part load the fans can be the dominant power source on the chiller itself and it may be better to reduce their speed and allow for higher compressor power consumption. It is that integration that is really important. An intelligent system recognises where to expend the energy and where to save the energy. Evaporators in stores is a good example. Probably more work needs to be done on looking at ice build-up within stores - but as long as we keep the doors shut, we won't have the ice.

Graeme Maidment
Philipp's graph showed the balance between operating at different conditions and how the controls, potentially, could bring that all together.

Philipp Holzer
The problem that you have is 'off the shelf' compressor controls that also control the gas cooler fans, the high-pressure valves and also heat recovery don't really take that into account. What we need is bespoke controls that actually look at the specific condition and decide at what point a change is needed. We could say that we are meeting the heat demand up to 17 degrees C by pulling free heat away from the system and it's only then that we may consider it's more efficient at that particular stage to revert to heat pumps to top up that heat. Maybe, when we get to really low ambient conditions, it might be more efficient to bring refrigeration back in and run the system transcritically at that point. Quite a lot of work is still being done around this topic to find the right strategy. Going back to

Nick's question about connected load, in terms of the refrigeration plant, it is important to try and establish how much heat is available. It is fine to say I've got a plant size that has 10% spare capacity and 9% utilisation at design conditions. But actually, how much connected load is there at one side? How much heat is available? I've been very fortunate to be able to look at that data from our monitoring system. It's based on very generic principles but it's better than just estimating there is maybe 50% that we could use. It's much better to rely on the actual dataset, and this is really important. One of the interesting things was that when we adjusted our design condition for a while to 35 degrees for our refrigeration plant from 32 degrees, and I looked at data from a site in Yorkshire and you hit 30 degrees ambient for just four hours a year. It would be easy to take a generic approach and to design everything for 35 degrees but the reality is we do need to look at this it a bit more carefully.

David Paget
Have you considered using phase change material for thermal storage to load shift or load balancing so you can make use of higher COPs and not have to run trans critically for so long?

Philipp Holzer
I have not. I have taken part in conferences about thermal storage recently and that's something that we need to consider. The next step, after looking at the data and trying to determine a strategy to look at false load. Certainly I expect that thermal storage is something that will form part of the future strategy.

Rob Lamb
Certainly food factories are the area where we have looked at thermal storage for storing heat in an industrial application because it meets both the supply and demand. You can supply the heat generated from the cooling process but do you need it all the time. Adding storage provides a way of reserving the heat for later use. Typically for a cooling process, eg spiral freezers, blast freezers or a food facility, it's when that's finished its operation and you've got to clean it down that you've got hot water. So having that type of storage and that type of phase change makes a lot of sense. Don't throw the heat away from your refrigeration system, store it to the point that you need it. Varying head pressure is also important. Run at a higher head pressure if you need to generate heat but only run part of the plant at a higher head pressure to generate the heat needed and allow the rest to run at ambient conditions if possible. That's one of the other things to think about. How much heat do you actually need? Typically you don't need all the heat that the facility is actually rejecting and you can portion a part of your system off to generate the heat that you want and run the rest of it in a more efficient, traditional cooling mode.

Graeme Fox
Part of the problem is that when the heat is actually being rejected is not always at a time when you need to supply that heat. That's why thermal storage is so important. You need to, as you say, not throw out the heat when you are actually producing it.

Graeme Maidment
Good point. Further questions from the room.

Martin Craxton
There has been a fantastic array of presentations tonight.
One of the issues that I am concerned about is balancing heating and cooling and as you mentioned, not throwing away that waste heat. There was an IOR presentation earlier on this year about a district heating cooling system. That seems to be the way forward. Using low grade heat pumped around in cities. I'm involved with data centres, and I think it's a shame that data centres are being sited out of town when, in fact, that waste heat could be used in a low grade heat network - instead of having air to water you could have water to water heat pumps, ground source heat pumps etc. I don't think there's basically room for all these air source heat pumps outside everybody's houses or blocks of flats.

Graeme Fox
I went to Stockholm about 4 years ago and toured the heating plant there. It was an old power station that they had converted into a heat pump, and it was absolutely phenomenal. They're recovering megawatts of heat out of the city's sewage system alone; let alone the heat pumps and they're just topping up that heat network. Personally, I think that heat networks are absolutely the future. The biggest problem we have is the mindset in the UK. Because we have been used to having our own heating system at a very localised level, in our own house, it's a very different mindset to effectively handing over control of the supply to a third party. We do have some of these in the UK, but not nearly enough of them. That is a big mindset change that we have to have to try and get across, particularly to housing developers. These data centres are ejecting huge amounts of heat and why are we throwing that heat away? If we have the thermal storage, we can capture that heat and put it into a district heat network system. It's all about the integration.

Rob Lamb
I think just a further point on that is that it is the infrastructure that is the key thing. It's national level thinking and its government thinking as well. Going back to the late 80s and early 90s we didn't have any problems digging up roads to fit cable tv into your house, everybody wanted cable tv. It needs that type of thinking, to dig up the roads, fit heating pipes and away you go. We're talking about a lot of energy, and we should be using it usefully.

Matt Wenger
My thesis was on wate heat recovery from data centres. In the latest government Heating and Buildings Strategy there is mention of a possible new efficiency rate for buildings that are consuming a higher rate of energy. Is there any chance this could push businesses towards waste heat recovery from data centres as a viable option in the UK? It seems to be happening outside of the UK, but we are really struggling to get it here.

Rob Lamb
I think this is desperately needed. It goes back to that question of infrastructure, because, where the heat sources are located is quite a distance, typically, to where you could use it. If it's in a city, for example, Birmingham, that could be a lot easier. But again, it's digging up the roads. It's the infrastructure. It is setting that requirement to utilize that waste heat. What I would say about data centres is that the opportunity for free cooling took a long time

to get integrated. As we move forward with data centres, we're removing the cooling from them in many cases, which is a good thing. It really does need some thinking about the rest of that heat and how do we use it? Where do we put it? Getting that community approach. Let's take that heat from our data centre and share it with somebody else, either next door or it could be a kilometre or two away through some sort of centralized heating system. We need policies that would move this forward, whether it's through the waste heat from a data centre or the energy being dissipated, let's use it somewhere else, through a heat pump to heat some buildings.

Mike Creamer
My thoughts have been very strong about the heat pump situation with the government now wanting to encourage many in a domestic environment and householders to move away from boilers. The manufacturers of such machines are producing phenomenal COPs now. Is the government focussed on these heat pumps being air to water on the basis that they won't be producing high enough temperature hot water that would suit the radiators in houses? I have solar tubes on my house generating hot water and I'm thinking how can I improve the system I have already got? Should I go to an air to water heat pump where I'm going to get high COP during the course of the day, but I want that storage for heat energy to be released at night for things in the house such as water and, of course, that means huge tanks. So, is the government thinking of air to water heat pumps only and what about the problem with the old housing stock and existing radiators?

Graeme Fox
Yes, they are generally thinking of air to water because they think that that's all heat pumps are! The rating level for new builds is B whereas most existing houses are D. That's what the problem is. That's why Insulate Britain are talking about putting all this insulation into homes ready to increase ventilation if we are going to have any chance of those air to water systems being retrofitted and being useful. But it's not just that. We have had this tradition of heating being installed using gas boilers to radiator systems using microbore tubing. So, it's not just a case of upsizing radiators, you actually need to strip out whole pipe system and put in 15mm pipe because you will not get the right flow rates for low temperature heating and sufficient kilowatts to heat the home. This is why we hear all these negative stories in the press about cost of installing a heat pump, it's because they're not just talking about the heat pump, they're talking about all the additional work that householders are having to do when you are applying it. That is a huge problem.

Mike Creamer
A reverse cycle air source heat pump is a great way getting instantaneous heating for the whole building or the house in no time but I because I could switch that into reverse and have 5 or 6 days of air conditioning in summer, it wouldn't be eligible, for a grant for example?

Graeme Fox
Yes, that is one of the big problems. We used to have a situation a few years ago where you had the enhanced capital allowances list and basically any of these reverse cycle air conditioning units were on that list with VAT discounts. Then, a few years ago, it became difficult to get the allowances, and the reason was specifically to avoid the use of air

conditioning units. I finally got to meet with HMRC last week and I was able to explain to them that whilst the air source heat pump can provide cooling they will primarily be used as heating systems. The questions that were coming back to me showed a lack understanding of the technology - systems are so much more efficient now they use refrigeration technology not resistance electric heaters to boost temperatures. But this is the problem, that we have policy makers who are talking about equipment that is 30 years old – the key thing is that the technology and the industry has moved on massively.

Graeme Maidment
There is a BEIS program I mentioned earlier on called Heat Pump Ready that will provide £70million funding for heat pumps and the development of heat pumps. I don't think cooling at the same time as heating will be off the table, the key is actually how you get heat pumps into homes. Do have a look at the program. If your business wants to bid to develop a heat pump you might get a grant to do that.

Graeme Fox
I think it's a really good point because what I am talking about is a conversation that I had two years ago but actually the conversation with Government has moved on and I think they are starting to understand that ventilation and air conditioning are a necessity far more often. So, yes, the conversation has moved on.

Mike Creamer
What I find interesting though is if you put in an air-to-air heat pump you still need a reversing valve to defrost it and so you can still switch it over to air conditioning for those few days of the year.

Graeme Maidment
In conclusion I'd like to reflect that over the past two years, even though we haven't got together, the discussion and the debate has moved on tremendously. It's very different than it was - we weren't talking about net zero but net zero is now fundamental to what we are doing today. In the background we have all done a lot of work.

I have got five takeaways from today in summary:

1. The importance of integration
2. The need to benchmark performance
3. Measurement is essential, not just generic solutions
4. Skills are going to be critical to achieving success
5. And the final one is collaboration - we need to continue to work together and share expertise and experience

Please join me in thanking our speakers once again.

IOR.org.uk
REFRIGERATION AIR CONDITIONING HEAT PUMPS

The knowledge hub for refrigeration, air conditioning and heat pumps

History of Flammable Refrigerants

Daniel Colbourne Ph.D. FInstR

Why you should attend

- By analysing the history of the use of flammables this paper will give a comprehensive background to the when, why, how and where of their use.
- It explores their role in the emergence of mechanical refrigeration, how they have supported the implementation of the Montreal Protocol and how the range of applications is now expanding.
- It also explains how different flammability classifications and safety requirements have emerged.

Summary

Refrigeration history is interspersed with the use of flammable refrigerants. During the 19th century, many of the early inventions employed flammable refrigerants. Until the time when CFCs became commercialised in the 1930s about a quarter of smaller systems used a variety of different flammable refrigerants, including ethyl ether (R610), methyl chloride (R40), ethyl chloride (R160), methylene chloride (R30), methyl formate (R611) and isobutane (R600a) and the majority of industrial/commercial systems used ammonia (R717). Whilst most of the refrigerants for smaller systems became substituted with CFCs, this was primarily due to mitigation of toxicity hazards, rather than ignition risks. As the Montreal Protocol came into force there were numerous studies assessing alternatives, including flammable HFCs and HCs. Initially, the domestic appliance sector briefly opted for R134a, but following the Greenpeace "Greenfreeze" campaign, virtually all European manufacturers adopted R600a. From the early 1990s, HCs became used more widely in small commercial and intermittently in other subsectors. In parallel, refrigeration safety standards were revised and amended to reflect the use of this technology. A decade or two later, safety standards were modified again to include the 2L lower flammability safety classification, igniting the commercialisation of flammable HFCs; primarily the introduction of tetrafluoropropene (R1234yf) in car AC systems and then difluoromethane (R32) in air conditioners. This paper traces the use of flammable refrigerants through textbooks, technical articles and patents, from the middle of the 18th century to the present day, breaking it into four "eras", mirroring the extent of use of flammables at the time.

1 Introduction

Refrigerants[1] – or "calorific media" (Siebel, 1918) – that are flammable[2] have interspersed refrigeration history. With the emergence of mechanical refrigeration, they were used simply because they were available and they worked well. Following the commercialisation of halocarbon refrigerants (chlorofluorocarbons, CFCs, hydrochlorofluorocarbons, HCFCs, etc.) they were set aside because they were less convenient; flammability of refrigerants was not a major issue, more of an inconvenience. Conversely, it was toxicity of many of the refrigerants – flammable and non – that posed the greatest concern. But as concerns over climate change/ global warming have become more prominent, the industry has little choice – by both environmental legislation but also chemistry – but to revert back to flammables for many applications. Furthermore it will be argued that whilst the use of a flammable refrigerants may instil resistance in some stakeholders, it is in principle a characteristic that is readily handled by good design and adherence to rules. Often exaggerated concerns and polarisation between refrigerant types, largely a hangover from earlier eras, is now complicating the wider use of flammable refrigerants.

Why should anyone be interested in the history of flammable refrigerants? Is the choice of refrigerant in the days of old of relevance today? It is hoped that a more intimate appreciation of the use and transition away

1 All refrigerants mentioned more than once are listed in Appendix: Table 4, along with safety and environmental data.

2 Throughout the literature, both terms "flammable" and "inflammable" (as well as "explosive") are used. Clarification on the former terms is useful. "Flammable" and "inflammable" both mean "able to burn". Instead of the prefix "in" denoting the negative, "inflammable" originates from the Latin "inflammare", meaning "to cause to catch fire." Whereas "flammable" is from the Latin verb "flammare" meaning "to catch fire". Initially, "inflammable" was adopted into English during the early 1600s, whereas "flammable" from "flammare" was coined in the early 1800s. Consequently, there are two synonyms that appear to be antonyms. Hereafter, "flammable" will be used. ("Inexplosive" is always the antonym to "explosive".)

Presented before the IOR on 2 December 2021

from and back to flammable refrigerants will provide a more dispassionate acceptance of the technology. For this, the chronology of flammable refrigerants is divided into four "eras":

- Unfolding of (flammables in) mechanical refrigeration – 1700s to 1930s
- The flammable interlude – 1930s to late 1980s
- Re-emergence of flammables – late 1980s to c. 2010
- The explosive era – c. 2010s onwards

These eras broadly reflect a level and extent of activity involving flammable refrigerants, but also the available (literary) sources, amount of information and degree of interest in such refrigerants. They are vaguely consistent with the "epochs" of Thévenot (1979) and the "generations" of Calm (2012).

As with any discussion on choice of refrigerants, it is apt to mention all the usual criteria: performance, cost, environmental impact, material suitability, etc. However, with a historical perspective, the author would also recommend the century-old equivalent in the article of Matthews (1920), which provides a entertaining "Benny-Hill-esque" overview of the various HC-based and other refrigerants. Regardless, the concluding statement is "Technically there are as many refrigerants, therefore, as there are different substances less one, some of which possess advantages over other, but none of which have all the advantages of the theoretically perfect refrigerant." The essence of this is recurrent throughout almost all publications presently drawn upon, including the recent Guidance Note 37 (IOR, 2021).

2 Unfolding of (flammables in) mechanical refrigeration – 1700s to 1930s

The unfolding of refrigeration technology appeared to be fairly slow; the occasional invention or demonstration here or there. But the idea slowly gained pace towards the second half of the 1800s. Synthetic chemistry began in the mid-1800s, but because refrigeration was still an embryonic technology, the new chemists of the period would not have chosen to seek out new refrigerants as a matter of priority. Therefore, refrigeration impresarios (as they apparently were) had to rely on any fluids that were available and which worked (Calm, 2012), regardless of flammability or any other characteristics. Similarly, there were almost no books, no refrigeration technology journals or magazines from this time, so the only available information is from patents and the occasional report.

2.1 The first flammable refrigerant

It appears that the first refrigerant was in fact flammable – ethyl ether (R610) – which coincided with the first mechanical refrigeration system. This was demonstrated by the legendary William Cullen (1710 – 1790), born in Lanarkshire, Scotland. His training and career appeared fairly diverse; he attended arts classes, a medical apprenticeship, was a ship's surgeon, an apothecary, had a medical practice, lectured in medicine, was the Chair of Medicine and lectured on chemistry at Glasgow and Edinburgh Universities. He spent many years formulating his own theory of heat and combustion and was referred to as "the true commencer of the study of scientific chemistry in Great Britain", although there seems to be considerable disagreement as to his contribution to the progression of chemistry as a science. For example, "Cullen wrote very little on chemistry and made no discovery" and his "hopes for drawing generalisations from a mass of miscellaneous information, were not novel" (Doyle, 2021).

Nevertheless, Cullen was surely remarkable in that at the University of Glasgow in 1748, he was the first to demonstrate artificial refrigeration. And it used a flammable refrigerant. In his demonstration, Cullen used a pump to create a small vacuum over a container of R610. When the R610 began to boil, it absorbed the heat from the container's surroundings, causing it to cool. This was recorded in his only published chemistry-related paper, "Of the Cold produced by Evaporating Fluids, and of some other Means of producing Cold". At least from the perspective of anyone who relies upon artificial refrigeration, this was surely a major discovery!

Coincidentally, whilst philosophising on heat and combustion, Cullen made a significant contribution to extending the current (at the time) theory of flammability: phlogiston theory. Phlogiston theory originated in 1667 from the alchemist and physician J. J. Becher (later formalised by G. E. Stahl) and postulated the existence of a fire-like element called phlogiston. This was contained within combustible bodies and released during combustion. Cullen's contribution was that of "mephitic air" that held out hopes for the physical isolation of phlogiston. The existence of phlogiston was based on empirical observations; the transfer and decomposition of phlogiston could be described by chemical explanation (Taylor, 2006). Although a number of chemists – notably Joseph Priestley, one of the discoverers of oxygen ("dephlogisticated air") and incidentally who discovered R717 ("alkaline air") – tried to retain some form of the phlogiston theory, by around 1800 it had been superseded by conventional combustion theory.

Cullen had a series of successors to (Nairne, Leslie, Vallance, etc.) who adapted and developed his ideas for mechanical refrigeration, often also using R610 (Thévenot, 1979).

2.2 Incendiary inventions

Over the 19th century there were a series of successive developments and inventions related to refrigeration technology, many of which utilised flammable refrigerants. These are widely described in books such as the legendary Thévenot (1979), Gantz (2015), Rees (2016) and the exuberant book of Wilson (1979).

In 1805, Oliver Evans (1755 – 1819; USA) designed a refrigeration machine, which was intended to use R610. Whilst he is often called the inventor of the refrigerator, he never actually built it. Apparently taking the ideas of Evans, Jacob Perkins (1766 –1849; USA/GB) obtained a patent in 1835 for "Improvements in the apparatus and means for producing ice and in cooling fluids". In this patent it states "I chiefly recommend [ethyl] ether as the material to be evaporated" and "I would have it understood that I do not claim the cooling of fluids by the evaporation of [ethyl] ether, or other volatile fluids; neither do I confine myself to the use of [ethyl] ether, though I prefer that fluid both as to its cost and its property of evaporating at exceedingly low temperatures." Such favourable views of flammable refrigerants run throughout the history of refrigeration.

About the same time, John Hague (1781 – 1857; GB) collaborated with Perkins to construct the machine, using R610. Apparently the first "compression expansion heat pump", using R610 was built by Hague and Perkins (Grace's, 2021), although there may be confusion over their hot water heating system.

A few years later in 1850, Alexander Twining (1801 – 1884; GB/USA) was granted a British patent for a commercial refrigeration machine for ice making (in the name of James Kingsford), which utilised R610 as the refrigerant. Moments later in 1855, James Harrison (1816 – 1893; GB/Australia) received a patent for a refrigeration system. The British patent 747 referred to R610, R717 or (ethyl) alcohol as the refrigerant. Daniel Siebe (1830 – 1866; GB) then opened a factory, achieving success in the 1860s and 70s with dimethyl ether (RE170) machines, based Harrison's system (Pearson, 2018). Away from Europe, Daniel Holden (1837 – 1924; USA) began building ice machines in 1870s with "chimogene" (Roberts, 2015).

About the same time, developments were also progressing in Europe, where Ferdinand Carré (1824 – 1900; France), whilst known for his development of the R717-water absorption system, also developed an R610 compressor around 1857. Also, Charles Tellier (1828 – 1913; France) developed an RE170 compressor in the

1860s and refrigerating plants both for static applications (factories, breweries) and ships for the transportation of perishable foods. German engineer Carl von Linde (1842 – 1934; Germany), patented the process of liquifying gas in 1876. His first machine, using R40, was completed in 1874 and tests proved its efficiency was double that of other existing equipment. Linde also developed a compressor for RE170 and an "improved" R717 machine in 1877, whilst David Boyle (1837-1891; USA) also developed one of the first R717 compressors in 1873.

There were many other characters that made significant discoveries and/or obtained numerous patents in refrigeration and related fields. As with recent times, not all of the major events and discoveries during the era of the unfolding of mechanical refrigeration involved flammable refrigerants, although many of them did.

From the late 1890s, refrigeration began to play a more prominent part in society; not only for the industrial and food storage sector but also within domestic settings. In 1899, A. T. Marshall received a US patent for a refrigerator, which used R717. The novel features were its automatic controls (expansion valve, a thermostat, a motor cut-off and condenser water regulator). In addition to some non-flammable options, R40 was used alongside R600a. Across Europe and North America, refrigeration units were being produced with flammable refrigerant; Figure 1 and Figure 2 show some examples.

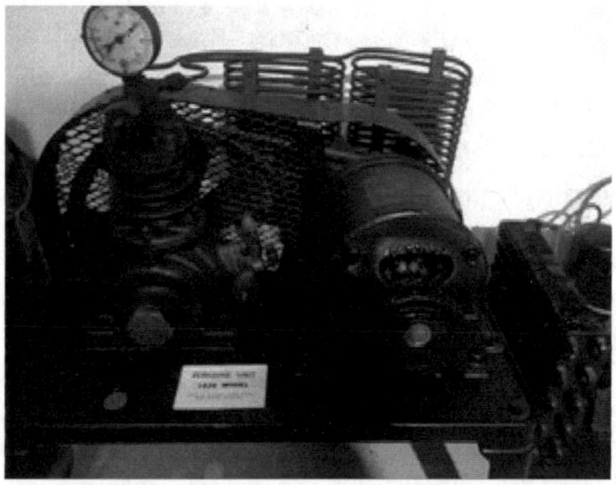

Figure 1: Zerozone R40 condensing unit from 1926 (Refrigeration Museum, Brighton, Michigan, USA)

Figure 2: Copeland R600a condensing unit from 1926 (Refrigeration Museum, Brighton, Michigan, USA)

Notably, during the early 1930s, Underwriters Laboratories (UL) produced a report examining the safety of various, including flammable, refrigerants (Nuckolls, 1933). Amongst other (non-flammables), it included R30, R40, bromomethane (R40B1), R160, ethyl bromide (R160B1), ethane (R170), propane (R290), butane (R600), methyl formate (R611), R717 and R1130(E). Not only does it provide flammability data such as ignition temperature, flammability limits, explosion pressure and flame propagation, it also describes toxicity of decomposition products. It concludes with comparative health hazards of refrigerants in the presence of flames and surfaces at high temperatures as well as fire and explosion hazards.

Despite many of the refrigerants in use towards the end of this era being flammable, the patents and later literature (trade journals, technical papers, textbooks) do not express concern (other than a philosophical context) or report on accidents due to flammability.

3 The flammable interlude – 1930s to late 1980s

During the interlude era, there is little discussion on the use of flammable refrigerants, except of course, for R717. As is well known, the use of many of the preceding fluids were substituted by the halocarbon refrigerants, so flammables faded from use in most applications. Arguably, the replacement halocarbons were not necessarily "better" refrigerants, but easier to use given there were fewer concerns to worry about. Indeed, it seems that the vast majority of accidents reported were related to the toxicity of various (flammable and non-flammable) refrigerants and not their flammability (Giunta, 2006). Regardless, as pointed out by McLinden and Huber (2020), the development and production of a fluid specifically for use as a refrigerant was undoubtedly a revolutionary development. Needless to say, it was found that this technology would ultimately have serious consequences (Kahlert, 2013). Although halocarbon refrigerants were commercialised in 1930s, most were initially discovered and synthesised between 1890s and 1920 by the Belgian chemist, Frédéric Swarts (1866 – 1940); he first prepared R12 in 1891 and many others followed (Banks and Tatlow, 1986; Kauffman, 1955).

3.1 Comburent textbooks

Very little of the literature covering this era refers to flammable refrigerants, so refrigeration textbooks have been studied to help form an impression. All of these textbooks introduce the choice of refrigerant in more or less the same way, offering the usual list of criteria to consider when selecting: performance parameters, non-corrosive, low toxicity, leaks should be easy to detect and locate and interestingly, while some state "definitely" non-flammable, they additionally say, but "preferably" non-explosive!

Throughout the interlude era there were many articles relating to thermophysical properties, heat transfer coefficient and pressure drop of fluids, including both HC and halogenated flammables, although none addressed application aspects.

Table 1: Flammable refrigerants mentioned in textbooks[3]

Source	Year	Methylene chloride (R30)	Methyl chloride (R40)	Ethyl chloride (R160)	Ethane (R170)	Propane (R290)	Butane (R600)	Isobutane (R600a)	Diethyl ether (R610)	Methyl formate (R611)	Ammonia (R717)	Dichloroethene (R1130(E))
Wallis-Tayler	1895								X		X	
Siebel	1918		X	X		X	X		X		X	
Hull	1927		X	X		X	X	X	X		X	
MacIntire	1928	X	X	X	(x)	(x)	(x)				X	
Moyer & Fittz	1932	X	X	X	X	X	X	X	X		X	X
Crawhill & Lentaigne	1934		X	X							X	
Althouse & Turnquist	1943		X					X		X	X	
Althouse & Turnquist	1944	X	X	X				X		X	X	
Lewis	1943		X	X				X		X	X	
Bäckström	1946		X	X				X			X	
Jordan & Priester	1948	X	X	X	(x)	(x)	(x)	X		X	X	X
Wostrel & Praetz	1948	X	X	(x)		(x)	(x)	X	(x)	X	X	(x)
BS1608	1949		X									
Collacott	1950		X								X	
BS 1725	1951		X								X	
Nelson	1953	X	X	(x)	X	X	X	X		(x)	X	
Sparks & DiLilio	1959										(x)	
Althouse & Turnquist	1960	X	X	X	(x)	(x)	(x)	X		X	X	
Breddy	1961		X								X	
Dossat	1961	X	X		X	X	X				X	
Fidler	1965		(x)								X	
Anderson & Herbert	1971		X								X	
Reed	1974		X								X	
Meacock	1979	(x)	X	(x)	X	X	X	(x)	(x)	(x)	X	(x)
Althouse et al.	1982		(x)	(x)	(x)	(x)	(x)	(x)		(x)	X	(x)
Gosney	1982		X		(x)	(x)					X	
Stoecker & Jones	1982		X		(x)	(x)					X	
Jones	1985										X	
Boast	1986										X	
Trott	1989										X	
Dossat	1991	(x)	(x)	(x)	(x)	(x)	(x)	(x)	(x)	(x)	X	(x)

(x) means listed but not discussed

Several interesting observations can be drawn from Table 1:

- R717 is consistently prominent from 1895 until the 1990s.
- Whilst R610 was pivotal to the development of refrigeration systems, it seems to have fallen from interest from the early 1930s onwards.
- Ethyl chloride (R160), R611, R30 and even R600a remained refrigerants of interest into the mid-1900s.

---

3 Books that give no mention of flammables have not been listed.

- Methyl chloride (R40)[4] drew much attention until the 1980s.
- Dichloroethene (R1130(E)), R600, propane (R290) and R170 seem to attract limited attention throughout the 100 years covered.

Hull (1927) mentions that of the approximately 500,000 domestic refrigerating machines in the USA, less than a quarter of them use flammables such as R40, R160, R600, R600a, R717, R290 and R610. It also states that R717 is used in more than 90 per cent of the larger or commercial refrigerating plants. R610 has some use in small hand operated machines which are manufactured in Europe and sold in the tropics. However, "a non-inflammable refrigerant is preferred in order to prevent danger in case of a gas leak in the refrigerating system in a home and also to prevent danger in case of fire." On R160, Hull remarks: "A certain quality of ethyl chloride has been produced in England which is claimed to be non-inflammable. This result is obtained by the addition of a certain amount of methyl bromide."

Bäckström (1946) estimates that R40 accounts for about 5% of refrigerant use in most sectors and R160 at about 2%, with R717 being about 10% across all sectors except industrial at 80%. The remaining are halocarbons. R600a was used in domestic refrigerators up until about 1933 (Jordan and Priester, 1948; Palm, 2018).

On R40, Lewis (1943) says "there is little danger of explosions, as with the amount used in an ordinary household refrigerator, if mixed with the air in the average kitchen would not make a mixture nearly strong enough to explode. It should be kept away from flames however." But "methyl formate is the refrigerant used by General Electric Co. in some of their refrigerators. ... It is anaesthetic if inhaled, and, is highly inflammable." And R717 is "somewhat inflammable and with the proper mixture of air will form an explosive mixture." However, whilst there are two pages on R600a, there is not a single mention of its flammability. Conversely, Jordan and Priester (1948) are somewhat unique in that they write in quite some detail, expressing concern over use of flammable refrigerants.

Interestingly, Siebel (1918) uses some antiquated terms, such as referring to "chimogene" – possibly a variation of "cymnogene", used elsewhere – which is probably a blend of R290 and R600 or "very volatile liquids derived from petroleum". It also refers to "rhigolene" (apparently a mixture of HCs with a normal boiling point of -9°C) and "cryogene" which is probably a mixture rich in R170 and/or methane (R50).

In terms of information on flammable refrigerants, MacIntire (1928) is somewhat of an exception amongst most, in that a detailed table of flammability characteristics is included, covering flammability limits, explosion limits, AITs, range of maximum explosion pressures, time to develop explosion pressure and "rates of diffusion". Conversely, within Moyer and Fittz (1932), despite detailed sections on R290, R170 and R600, the topic of flammability is not even mentioned; any safety discussion is related only to toxicity. As with numerous authors, Siebel (1918) seems to have a serious affection for R717, particularly illustrated with the section entitled "Ammonia not explosive", where he goes on to explain that R717 is only as explosive as water (under pressure).

3.2 Fulgurating flammables

Once non-flammable, lower toxicity halocarbons became established across most parts of the industry, the use of flammable refrigerants dwindled. However, there appeared to be a few instances of their use, when commercialised refrigerants were not suited or unavailable.

4 R40 was the fluid within counterfeit R134a that caused a series of explosions with reefer containers (UNEP, 2012)

One notable work was an ingenious invention of the seemingly eccentric German (residing in USA), Peter Schlumbohm (1897 – 1962). His concept was to use R290 in a "transitory cycle" for transport refrigeration, where it first passes through the evaporator and then as a motive fuel to the truck engine, shown in Figure 3 (Schlumbohm, 1939). Schlumbohm had exchanged the inheritance of his family's chemical company with funding of a lifelong education, although "the improvement of refrigeration through chemical, mechanical, and engineered processes was Schlumbohm's perennial interest throughout his working life"; he considered refrigeration to be one of the most important and necessary scientific fields (Anon, 1949). He had over 100 patents related to refrigeration, including the transitory R290 system (Schlumbohm, 1937).

Figure 3: Photograph of R290-driven refrigerated truck and fuel system (Schlumbohm, 1939) (©ASHRAE www.ashrae.org Refrigerating Engineering, Economic Application of Air Conditioning and Refrigeration, Vol 37, No 4, Apr 1939)

From the middle of the 19th century, there was far less discussion on flammables. WWII had certain impacts on refrigerant use, though. Since in Europe at least, halocarbons were diverted for military use, it resulted in a return to the use of certain flammables in household refrigeration (Palm, 2018). Apparently, a major concern was that bombing could result in major releases of refrigerants, such as R40, R717 and some non-flammables, with catastrophic impacts on building occupants (McNally, 1943). Stonebanks (1949) reporting on the need to operate low temperature systems during WWII in absence of R22, detailed extensive successful trials carried out with R290. He remarks: "Under 'standard ton' conditions, propane compares favourably with other refrigerants in more common use." and "The main disadvantage is, of course, its inflammability and though in certain cases this might be a source of danger our experience, taking reasonable care, has been entirely without untoward incident".

In Fidler (1965), chapter author Webb was apparently besotted by halogenated refrigerants: "with the rapid increase in the use of the halogenated refrigerants, there has been a more or less complete falling off of carbon dioxide, sulphur dioxide and methyl chloride, due either to very high pressures involved or a low critical temperature, or to toxicity, methyl chloride being an anaesthetic." There is little mention of other flammables, but reluctantly adds "The other classic, ammonia, is still being widely used and is likely to be so for some time to come."

Still, for most industrial applications R717 played a pivotal part throughout this era (Stoeker, 1998). Similarly, despite R290 (and R1270) not being used for smaller applications, its was retained for chemical and industrial process systems globally (Perry, 1984), as well as for liquefied gas shipping. For example, Howdens, Stal, etc. continued to produce a range of R290 compressors for large systems.

Although infrequent, there are reports of interesting work with HC mixtures during this era. Carr (1949) trialled tertiary blends of R290, R600 and R601. Haselden and Klimek (1957) and Klimek (1959) report on experiments of HC blends for the purposes of energy saving in heat pumps (HP). Other studies from 1939 and 1947 are also cited. Another article also evidently ahead of its time, looked at amongst others, blends of R290/R601 and R600/R601 as options for industrial HPs (Schnitzer, 1983).

Although the interlude era was dominated by the halocarbons, flammables were used occasionally – often for their favourable performance characteristics – and when they were, ignition risk was not considered a major issue; appropriate design of equipment and handling was all that was required.

4 Re-emergence of flammables – late 1980s to c. 2010

The era of re-emergence of flammables covers events within (some of the author's) "working" memory, coincidentally, from around the time of ratification of the Montreal Protocol. The Montreal Protocol sparked off a series of events: extensive probing of possible alternatives to ozone depleting substances (ODS), the Greenpeace Greenfreeze campaign, adoption of HCs by domestic refrigerator manufacturers, proliferation of HC to other equipment and in parallel, revision of safety standards and stakeholder studies on alternatives.

4.1 The Montreal Protocol

Under the framework of the 1985 Vienna Convention for the Protection of the Ozone Layer, the Montreal Protocol on Substances that Deplete the Ozone Layer was adopted on 15 September 1987 and entered into force in 1989. It aimed to phase out both the production and consumption of ODSs, including CFCs and latterly HCFCs. With the looming phase-down, industry began to rummage around for replacements leading to a torrent of studies examining numerous substances as potential alternatives, many of which included flammables.

One of the earliest and perceptive studies, of McLinden and Didion (1987), took a "molecular approach" to screening for possible alternatives. Despite starting with "must be nonflammable", they conclude that "some compromise with the traditional criteria [including flammability] is inevitable." Soon afterwards, Kuijpers et al. (1988) appraised various flammables, including R152a, R161, RE170, cyclopropane (RC270), R290, R600a, propadiene (R2250) and propyne (R2250b); the latter two having two unsaturated bonds. Whilst it recommended R134a, secondary proposals were R152a, RE170, RC270 and a blend of R290 and R600a. Similarly, Jürgensen (1992) presented test results with domestic refrigerators, using not only R600a, but also R290, R600 and blends. Missenden et al. (1990) and Missenden and James (1992) reported an insightful study on R290 in refrigerators, covering comparative performance, combustion safety tests and deliberation of safety standards. Bodio et al. (1993) studied the performance of R290/R600 mixtures in domestic refrigerators and concluded that the performance is at least the same or better than R12. Also for domestic refrigeration, Snelson et al. (1991) found R152a to give considerably better performance than R12 and R134a, whilst Devotta and Gopichand (1992) favoured RC270 and R152a. Kuijpers et al. (1991) further reported extensively on tests with R152a and Shiflett et al. (1992) positively evaluated various flammable HFCs. Others (e.g., Sanvordenker, 1992) challenged the advantages of R152a, although Spatz (1991) stated that the flammability of R152a would either eliminate it from consideration or impact costs by requiring additional safety controls. Less familiar flammables such as blends of R290 and RC270 were found to yield lowest energy consumption (Camporese et al., 1991). Whilst R152a was looking like a promising option, it seemed to fade from interest.

Given the apparent interest in flammables, there were early risk assessments on domestic refrigerators (Pelto and Harris, 1990) and for wider application (Dierckx and Berghmans, 1993). Richard and Shankland (1991) published an impressively comprehensive study of ignition characteristics of numerous flammable refrigerants. It reported on flammable limits of more than 20 fluids, in many cases using a variety of ignition sources. There were even detailed studies examining the explosion risks of HCs in hermetic compressors (Zgliczynski and Sansalvadore, 1994).

Whilst domestic refrigeration seemed to be a major focus for alternatives at the time, flammable alternatives for HPs was also an area of notable interest at the time. For example, Kruse and Hesse (1988) proposed R142b and R152a for HPs, whilst Pannock et al. (1992) looked at a mixture of R32/RI52a. Bivens (1991) considered various flammables for HPs, including R290 and R32 but errs towards non-flammable R32 mixtures. Other applications were also considered for flammables. For transport refrigeration, Boldrin et al. (1991) examined several various refrigerants including R290, with regards system performance as well as toxicity, environmental impacts and costs; of course, R290 was found to provide highest efficiency. Treadwell (1991) favourably addressed safety aspects of R290 for a packaged AC unit and presented test results, identifying numerous cost, environmental, efficiency and performance benefits. In general, Kramer (1991) argued that whilst R290 is flammable, it is highly effective, readily available, compatible with standard lubricants and with proper precautions and approvals, can be substituted for R22 and R502.

Surprisingly, R32 seems not to have been considered as a single component (within the literature) until 1991. Before 1995, there were almost no studies looking at (pure) R32; only mixtures. Powell (1991) contemplated R32, but then erred towards mixing it with R134a to avoid flammability.

4.2 The "Greenfreeze" detonation

Whilst these various debates were ongoing, the domestic appliance sector decided upon R134a as the replacement for R12 – despite a GWP in the 1000s and numerous studies demonstrating other alternatives were more favourable. However, the environmental campaign network – Greenpeace – who had been involved in the earlier movement against ODSs, became concerned about substitution of R12 with R134a. In the early 1990s, after aerosols, the domestic refrigerator had become a symbol for CFCs. In order to address the CFC and subsequent HFC problem, Greenpeace (Germany) tried developing its own CFC-free refrigerators, such as with a Stirling engine. These were, however, all failures since efficiency was poor compared to existing CFC models. During this process, an associate from the Dortmund Hygiene Institute, Dr Preisendanz, reported on his effective use of HCs; he with Wolo Lohbeck and Harald Zindler of Greenpeace set about introducing HCs to domestic refrigeration.

Initially, the mainstream West German (at the time) manufacturers rejected HC alternatives. However, a struggling company in the former East German Democratic Republic (GDR), Deutsche Kälte- und Kraftmaschinen (DKK) Scharfenstein eventually expressed interest. (At their pinnacle in the 1980s, they had 5300 employees, producing >1 million appliances a year, being sold as discount products in the West and also across the Eastern Bloc.) Although, as with other manufacturers, they has also switched to R134a. From March 1992, Lohbeck, Zindler and Preisendanz visited DKK on several occasions to discuss the possibility of switching to HCs. (Interestingly, DKK found records that they had already worked with HCs in the 1930s.) Subsequently, Greenpeace concluded an agreement with the factory to build ten HC refrigerators and organised a press conference. In the meantime, the Treuhandanstalt (or "Trust agency") – an agency established by the West German government to oversee privatisation of East German enterprises – was planning to act upon DKK, which would have blocked any possibility to switch to HCs. The Treuhand director travelled to Scharfenstein to prohibit the press conference, which failed, because there were 150 journalists already present and as a consequence the Treuhand ended up guaranteeing continuation of DKK. The factory was renamed Foron in January 1993.

An additional advantage of DKK/Foron was that they also manufactured compressors; given the objections of the appliance manufacturers, it was unlikely that any compressor manufacturer would readily supply HC models for the project. A blend of R290 and R600a (named "Dortmunder Mischung" or Dortmund Mixture after the institute where it had been conceived) was chosen as the refrigerant so that minimal changes to the compressor would be required thus enabling rapid transition. This was crucial for the success of Greenfreeze. With all the bedlam surrounding DKK/Foron, instead of the ten test fridges, they directly prepared for mass production. The first series models of Greenfreeze Foron fridges were shown at Domotechnika in Cologne in February 1993 and the first series of Greenfreeze fridges (model KT135R) were completed on 15th March 1993 (Figure 4). In the meantime, having ceased their anti-HC campaign, the other leading manufacturers at the time also presented their first HC prototypes.

Figure 4: The first HC fridge from DKK/Foron (Sabine Vielmo/Greenpeace) (left) and on display (Manfred Scharnberg/ Greenpeace) (right)

As with today, the principal objection to using HCs was the flammability risk. Greenpeace and Foron were immediately attacked for the "explosion hazard" and the leading European manufacturers (AEG, Bauknecht, Bosch, Electrolux, Liebherr, Miele and Siemens) jointly wrote a letter in 1992 to the trade associations warning against a "new, allegedly CFC-free refrigerator". This letter, being unlawful according to the anti-competition rules, led to the manufacturers having to retract their objections and apologise. Regardless, videos appeared showing "German" refrigerators exploding. Despite such activities, the product standard IEC 60335-2-24 (IEC 335-2-24 at the time) was revised in tandem to allow up to 150 g of flammable refrigerant to be used, provided potential ignition sources on the appliance were eliminated. Anecdotally, there was never an accident with these DKK/Foron refrigerators.

The existing standard (BS 3456-202.24: 1990 – the BSI implementation of IEC 335-2-24) did not prohibit flammables; in fact there were no requirements related to refrigerant safety. But for one exception: in Annex ZX, there were a list of special "national conditions" (deviations) and applicable to the UK only, it was stated: "The refrigerant shall present no more hazard than those listed in the following table, except that ammonia shall not be used in compression type appliances." The list included R22, R12, R11, R115 and R502. The deviation was due to expire on 1st January 1991.

A standard that simply does not prohibit a flammable refrigerant is not as helpful as one that provides guidance, so for the fourth edition in 1997 it was revised and published, including a limit for the HC charge of 150 g[5] and associated measures for marking and safe electrical components. Revision of the standard was fundamental to success of R600a. Within a year or two, DKK/Foron, as did most European refrigerator manufacturers, switched to R600a. Hermetic compressor manufacturers also responded and Liebherr's "table top" fridge with an Aspera compressor was the first R600a product for 60 years.

It is evident that Greenfreeze opened up the possibility of extending the use of HCs to many other applications within the refrigeration, air conditioning and heat pump (RACHP) sector – today we see extensive use in commercial refrigeration, portable ACs, chillers and HPs. That R600a is used in more than one billion fridges (UNEP, 2018) gives confidence to other sectors. Not only was this an incredibly successful campaign for Greenpeace, but it paved the way for the wider introduction of HCs to other applications. Arguably, it may also have enabled a smoother introduction of flammable HFCs ("A2Ls"). Since HCs have negligible GWP, legislators saw HCs as viable alternatives to HFCs, so as legislation edged towards pushing for lower GWP, halocarbon producers developed more suitable alternatives and flammable HFCs were the most appropriate options on their table. Moreover, because HCs had become the flammability "punch-bag", it was not only easier to introduce A2Ls, but they could be offered as an antidote to "high flammability" HCs.

4.3 Greenfreeze babyboom

Undoubtedly, Greenfreeze generated a momentum for HC refrigerants. Globally (except in North America), more and more refrigerator manufacturers switched from CFCs or HFCs; by 2010 more than 50% of new production used R600a (UNEP, 2015). The initial conversions in Article 5 countries were carried out by implementing agencies, funded by the Montreal Protocol's Multilateral Fund. The first factory conversions, from R12 to R600a, was by GTZ (now GIZ) Proklima in China (Kelon and Haier) in 1994. This was followed by projects in India, Pakistan, even Cuba, and latterly Argentina and Brazil, again by GTZ, Infras and others. In a separate development, an Italian appliance manufacturer, De'Longhi, launched a range of portable and split ACs with R290 in 1994, as well as some Scandinavian and German HP manufacturers.

Correspondingly, numerous producers of HC refrigerants materialised worldwide: Ecozone, Elgas, Linde Gas, Musicool, Oz Technology, Settala Gas and Calor Gas, amongst (numerous) others, whilst Hychill/Esanty (Australia) in fact began even earlier in 1985. Some of these enterprises created significant impetus for the progression of HC refrigerants, working with refrigerator, commercial refrigeration, ACs and other manufacturers to extend application ranges.

There continued to be many studies focussing on HCs in various applications and with a surprisingly detailed treatment of safety aspects. Frehn (1993a and 1993b) argued that none of the leading options for HPs offer as many favourable characteristics as R290, as did Halozan (1994) and Eggen at al. (1994). Safety was further addressed by Granryd et al. (1994) including risk assessment approaches, whilst a more formalised quantitative risk assessment approach was described by van Gerwen and Jansen (1994).

Even Hickman (1993) posited that R290 and R717 are likely to receive fresh consideration in the turmoil created by changes. Maclaine-cross (1993; 1994) looked at the application of HCs to motor car air conditioning systems (MACS), concluding that HCs led to better efficiency and negligible flammability risk. At the same time, there

5 It is frequently asked "why 150 g?" The explanation is broadly, since a "typical European kitchen" is about 8 m^2 x 2.5 m high = 20 m^3 then using a 20% factor applied to the LFL of R600a gives: 20 × 0.2 × 43 = 172 g. Based on the charge that most manufacturers needed at the time for the majority of models, 150 g suited.

began widespread replacement of R12 (and latterly R134a) by HCs in MACS. Later, there were also studies looking at – and associated patents for – flammable refrigerants mixed with fluoroiodocarbons to inhibit flammability (e.g., Nimitz and Lankford, 1994). 25 years later this is being repeated in the form of mixtures such as R466A.

The attention on HCs prompted revision and drafting of further other standards, such as BS 4434:1995 and DIN 7003: 1995. With the wider interest in refrigerants such as HCs and R717, the International Institute of Refrigeration (IIF-IIR) held the first conference on natural refrigerants[6]. This was in Hannover in 1994 ("New Applications of Natural Working Fluids") and became the "Gustav Lorentzen Conference" at Oslo in 1998. Similarly, the trade association, Eurammon, which swiftly opted to encompass all natural refrigerants, was set up in 1996. "Market accelerator", Shecco began in 2000, later expanding their purview to also encourage faster adoption of flammables, R717 and HCs.

In 1998, the Greenfreeze campaign turned its attention to the major corporate sponsors of the 2000 Sydney Olympics – giant food and beverage companies such as Coca-Cola, McDonald's and Unilever, whose operations used millions of refrigeration systems worldwide. As the opening of the Sydney Olympics approached, Greenpeace took various initiatives to convince these sponsors to adhere to the "environmental guidelines" issued by the Olympic Committee. Subsequently, Coca-Cola, McDonald's and Unilever announced plans for a global phase-out of HFCs. This led to the Refrigerants, Naturally! Initiative, also encompassing other brands such as PepsiCo, Ikea, Heineken, Carlsberg and Red Bull (plus supporting participants, Greenpeace and UNEP). The momentum created by the global adoption of HCs (and CO_2) across these end users' millions of commercial refrigeration units led to wider acceptance of flammables in the retail sector. Manufacturers began large-scale production of hermetic R290 compressors and the global output of HC compressors outstripped that of R134a.

Beyond retail refrigeration appliances, numerous manufacturers were beginning to offer products with HCs: HPs, split and portable ACs, chillers, etc. However, whilst many system manufacturers were using R22 compressors, some prominent compressor producers were opposed to the use of flammables. With the entry into force of the European Pressure Equipment Directive (97/23/EC), this practice then became prohibitive; few compressor manufacturers offered HC-approved equivalents, thus the range of HC systems began to dwindle. Indeed, a further catastrophe for some AC manufacturers was the approval of the standard, IEC 60335-2-40 in 2003, specifying charge limits for flammables which effectively obstructed the use of HCs in many such systems (Colbourne et al., 2020). Whilst in Europe, the focus was more on new systems, a large retrofit/top-up market established throughout Asia/Australasia, North America and even the Middle East, particularly for the MACS sector. For instance, an analysis of samples taken from car AC systems in Thailand found about 25% contained HC only and a further 25% with R12 or R134a mixed with HC (UNDP, 2000). Maclaine-cross (2004) estimated in excess of 5 million cars using HCs in MACS, a number likely dwarfed by the numbers is China and other parts of Asia. Considering the astronomical number of "drop-ins" and the absence of reported incidents, in retrospect the scare stories (at the time) appear unjustified.

Although the use of HCs in RACHP equipment was in flux, there were numerous patents being filed (e.g., Lindgren, 2002), although similar mixtures were patented almost a century before (Seaman and Crawford, 1918). Not only were these patents originating from smaller companies, but also multinationals such as Exxon Mobil and even Du Pont (now Chemours) applied for HC refrigerant patents, presumably in anticipation of

6 The term "natural refrigerant" is apparently a contentious one. In the current context it is used to characterise fluids that are widely found in nature and which are regarded as sustainable. Principal natural refrigerants are hydrocarbons, ammonia, carbon dioxide, water and air.

increased acceptance and growth of HC refrigerants. Amongst these were several imaginative inventions such as a "hydraulic" refrigeration system where HC refrigerant is circulated with water, thus rendering a leak "non-flammable" (Rice et al., 1991), or adding an odorant such as orange blossom to provide a warning in the event of leakage. Indeed, later patents have included flammable unsaturated HFCs and HCs along with a variety of potential odorants: fresh lemon, cherry, cinnamon, peppermint, or orange peel (Minor, 2009).

With the growing adoption of HCs in certain applications, emotions flared. Whilst not clearly documented, there was a distinct polarisation of views between those backing flammable HCs and non-flammable HFCs. Meetings, symposia and conferences were often distinctly pro- or anti- or would have refrigerant evangelists spouting in favour of one or another. Once, after giving a presentation on flammables, an audience member looked the author in the eyes and exclaimed "You are trying to kill me, my wife and my children!"

5 Explosive era – c. 2010s onwards

From around 2010, there was an explosion in the study and use of flammable refrigerants, evidenced by the commercialisation of flammable HFCs. On one hand, R1234yf was beginning to become commonplace in MACS and soon R32 in static AC systems. On the other hand, a wide range of flammable HFCs and blends thereof were being developed and offered by refrigerant producers. Beyond domestic refrigeration, HCs became widely used in retail environments (and sporadically in large integral systems such as chillers), not only in Europe but also in North America, Asia, Southern Africa and Australasia. In particular, the decision under the Montreal Protocol in 2007 to accelerate HCFC phase-out heightened the need for suitable alternatives in Article 5 countries. Subsequently, there were numerous demonstration projects with a focus of natural refrigerants, executed by implementing agencies such as UNDP, UNIDO and GIZ Proklima. Example projects include HC chillers in Indonesia and R290 room ACs in China and India and commercial refrigeration using HCs in Southern Africa. One notable success was the introduction of R290 split ACs into India, by Godrej (Rajadhyaksha et al., 2015). The current output in excess of one million units. Indeed, it is curious that HCs have not been applied more widely for the split AC sector, given the (flammability) risk is hundreds of times lower than that of HC refrigerators (Colbourne and Suen, 2015).

5.1 The great inversion

Earliest mention of flammable HFCs seems to be around 1900 by Swarts (Kauffman, 1955) and unsaturated HFCs (fluorinated alkenes or hydrofluorooelfins, HFOs) around the 1940s; for example, Henne prepared R1234yf and R1132a in about 1945 (Sicard and Baker, 2020). McLinden and Huber (2020) provide a detailed description of the various research efforts during this period to identify viable molecules to address the ODS and GWP pressures. The possibility of commercialisation of unsaturated HFCs first appeared as a patent by Daikin for R1234yf and similar compounds ("fluids for heat transfer purposes") in 1990. This was closely followed by a Russian patent that included a greater number of unsaturated HFC molecules – R1216, R1225, R1243, R1252, R1261 as well as R1234 – and also mixtures with HCs (Nikolaevich et al., 1994). From there on, there are innumerable patents on such fluids, applied for by all the mainstream refrigerant producers. Despite decades of received wisdom that refrigerants shall not be flammable, environmental pressures largely forced the great inversion; gradually it was no longer heresy to view flammables as possible mainstream refrigerants.

In the early 2000s, preparatory work on the first European rules to control fluorocarbons, or "f-gases", began. With the anticipated MACS directive (2006/40/EC) likely to limit refrigerants to GWP<150, halocarbon producers were desperate to find a suitable alternative; MACS are a highly profitable application due to the huge numbers of systems and the high re-charge rate, representing more than a third of global refrigerant consumption. Numerous options were on the table including HCs, R744, R152a and several concoctions based around R1234yf. Eventually, pure R1234yf was opted for in the majority of instances and wider use progressed from the early 2010s. Extensive work and discussions is reflected by the industry projects, SAE CRP (2021). Whilst half of the debate revolved around performance, the other half was related to flammability issues (UBA, 2021), which – of all systems – seems irrational for MACS, given the extent of flammable fluids already used in cars (brake fluid, engine oil, petrol/LPG/LNG/hydrogen fuels, etc).

Another sector about to begin a shift to flammable HFCs was static air conditioning (ACs). Developments had been ongoing since the 1990s, but there were numerous studies assessing mixtures of R32 with non-flammable HFCs (such as R410A). From around 1993/1994, there were many studies looking at R32, the majority of which found significant improvements in performance with the pure substance (e.g., Yajima et al., 1994), so it is surprising that it was neglected for so long. The first R32 split AC was launched in 2012 (Daikin, 2021) and gradually more AC producers, initially from Japan but then other countries, followed suit.

Although, as implied above, the number of patents covering flammable refrigerants is mindboggling, a snapshot of them helps provide an impression of what portions of the industry is thinking with regards to the ongoing direction of flammable refrigerants. Patents cover both the fluid (molecule, production, etc.) and systems and components. Surprisingly, the number of patents associated with HC refrigerants is broadly comparable with those of fluorinated flammables. The study of NGBIPRI (2014) found that in Japan alone, of 810 patents, 619 are related to flammable HFC refrigerants and 284 for HCs. Internationally, of 1026 patent families (i.e., of the same invention but covering a number of countries), 780 are related to flammable HFCs and 395 to HCs. Many of these patents involving HCs were registered to halocarbon producers: DuPont (now Chemours), Arkema, Central Glass Co. (Ashahi Glass), Mexichem (Koura) and Honeywell. Of 426 patents for R1234yf alone, many are being disputed (in 2020) without a ruling (Booten et al., 2020; Anon, 2021). Roger (2015) uncovered 1370 patents relating to flammable refrigerants, of which nearly 1000 were for R32.

5.2 Flammability hazards in perspective

Of course, the introduction and commercialisation of flammable HFCs has involved discussion that has been to the detriment of higher flammability (HC) refrigerants, where HC flammability has been used to "cushion" the impact of lower flammability HFCs (e.g., EFCTC, 2021).

Nevertheless, it must be recognised that the flammability classification of refrigerants is not clear-cut, for instance, the well-known class A1 refrigerant (Table 2), R22 is flammable under pressure. Tsai et al. (2018) show that the explosion pressure can be up to 30 times that of the start pressure; ignition of R22-air mixture within a compressor at 10 bar can result in an overpressure of 300 bar. Indeed, many accidents have been reported, involving fatalities due to explosions of "non-flammable" and lower flammability refrigerants (e.g., Behera et al., 2017; Moga et al., 2021). These tend to happen during servicing (partly because air is more likely to be introduced into the system, but more so because persons are present in close proximity to the system), but such explosions have also been reported in absence of servicing. A simple search on the internet yields an extensive number of reported accidents – six examples found within a few moments of searching.[7]

7 https://www.youtube.com/watch?v=GxYJlR5Eajg

https://www.channelnewsasia.com/asia/man-killed-son-injured-air-con-compressor-explodes-seremban-881821; https://www.youtube.com/watch?v=iu4F8e1iJZ4; https://www.coolingpost.com/world-news/one-dead-two-injured-in-air-conditioning-explosion/; https://timesofindia.indiatimes.com/city/thane/thane-mechanic-dies-after-blast-in-ac-compressor/articleshow/62626750.cms; https://www.chiangmaicitylife.com/citynews/local/ac-explosion-led-severe-injuries/ ; https://www.coolingpost.com/world-news/second-man-dies-from-chiller-blast/

A study in Japan helped to elucidate the situation. Higashi et al. (2017) and Higashi et al. (2018) measured Diesel explosion pressure with refrigerant-air mixtures and various types of oils. They found that not only flammable refrigerants, R32 and R1234yf produced such explosions, but R410A and R22 as well. Depending upon the type of oil, the explosive range of R32 and R1234yf was up to between 1% to 50% in air. Thus, there is a large window of opportunity for creating an explosion when drawing an air-refrigerant mixture into the compressor. Conversely, measurements with R290 showed that under the same conditions, there was only a range of 0% to 2% where a Diesel explosion could occur, regardless of the oil type. In other words, the compressor needs to be drawing a 98% mixture of air with refrigerant for the possibility of ignition to occur; this is considerably more difficult. Thus, these severe accidents seen with non- or lower flammability refrigerants are almost impossible with HCs. Unfortunately, the ill-informed can easily jump to incorrect conclusions (Refcom, 2021). Further, whilst some external (to the system) ignition studies have found modest consequences with A2L refrigerants (Gandhi et al., 2017; Davis and Pagliaro, 2019), some experiments with A2L refrigerants and oil have shown surprising results that are seldom discussed (Kopylov et al., 2019).

Whilst HC refrigerants would undoubtedly lead to more severe consequences than A2L and A2 refrigerants, external to the system, this does not automatically mean that such consequences will regularly occur. Accidents have of course happened with HCs. Serious accidents have been very few and far between, but have invariably occurred as a result of people doing daft things; dropping HCs into very large systems (designed for A1s) without first eliminating sources of ignition (as safety standards demand) or carrying out brazing or electrical work in a confined, unventilated space. Small numbers of certain domestic refrigerator models exploded due to manufacturing faults, resulting in recalls and changes to the safety standard to avoid such problems with future models. However, national statistics demonstrate fire frequencies due to flammable refrigerants are minimal (DCLG, 2014), less than one-hundredth of accidents arising from electrical faults. Indeed, frequency of fires from refrigerators over the period that R600a supplanted R12 and R134a shows a gradual reduction (likely due to improvements in electrical components).

Over the preceding decade or so, flammable refrigerants had become more widespread and, in some sectors, prominent. Despite the often emotionally- or commercially-driven opposition to flammables, accidents are extremely rare and less frequent than electrical fires (in the same equipment) by orders of magnitude. Of course, as such events occur, safety standards are revisited and refined accordingly.

6 Safety standards and similar rules

Rules and regulations for handling flammable substances have been in place for decades and identification of requirements to enable safe application of flammable refrigerants is relatively easy (once one has become familiar with the principles and workings of these regulations). However, concluding upon practices specific to refrigeration systems, extracted from highly generic rules for any application of flammables relies upon subjective judgement. For this reason, industries prefer dedicated rules tailored and refined for their specific circumstances. Hence the evolution of safety standards.

Although in many countries, safety standards are not necessarily legal requirements, they somehow seem to be pivotal for the application of flammable refrigerants. For example, in the UK, in principle, a system must comply with the relevant parts of the Supply of Machinery (Safety) Regulations 2008, Dangerous Substances and Explosive Atmospheres Regulations 2002 (DSEAR) and Equipment and Protective Systems Intended for Use in Potentially Explosive Atmospheres Regulations 1996. However, this somehow presents difficulties, so the various rules must be interpreted and expressed as specific requirements for RACHP equipment. Nevertheless, safety standards somehow act as traffic lights to the use of flammable refrigerants; if a standard permits, the sector can adopt, if it prohibits (despite the regulation allowing), the industry shuns it.

Generally, major accidents have resulted in changes in rules (legislation, standards, etc.) so given the progress of time, this tightening of rules led to fewer accidents. As an illustration, as seen in Figure 5, the number of annual fatalities of employees reduced from about 5000 to 100 over the past century. Prior to 1974, health and safety (H&S) legislation was less inclusive (with regards to breadth of industries) and reporting less rigorous, thus the actual number of fatalities in earlier times was likely much higher. This demonstrates that the effectiveness of wider H&S legislation, rigour of enforcement and level of awareness greatly improved levels of safety over this period. This must similarly apply for safety standards RACHP to systems.

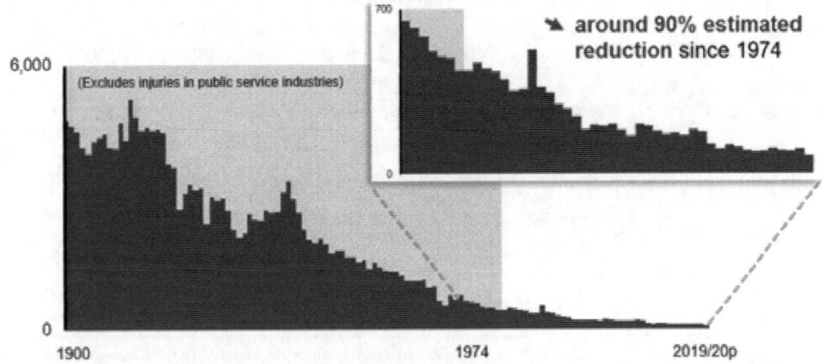

Figure 5: Number of fatal injuries to employees in Great Britain 1900-2019/20 (HSE, 2020)

In terms of flammability safety in general ("explosion protection"), the first notable contribution was made by Davy (1816), who developed an oil lamp for mines that prevented flame propagation by means of a close-meshed screen, which was also, incidentally, the first mechanical gas sensor. (Davy also reportedly was the first to discover R717 in 1807.) Later in 1884 the first "flameproof" enclosure, enclosing an electric switch, was patented by Theophilus Cad (British Patent 806). National rules seemed to lag these practical developments, though, with (one of) the first for explosion proof or "Ex" equipment being VDE 0170 in 1912 in Germany and BS 229 later in 1929 (Górny, 2013). Despite the not insignificant use of flammable refrigerants over this period, all of these developments were aimed at the mining industry. From the first Ex safety standard in 1957, the list has grown to nearly 50, reflecting the rise in knowledge and technologies applicable to explosion protection.

6.1 Safety classification

It is useful to consider fluid safety classifications. Most industries handling flammable gases use a system different to the alpha-numeric approach of the RACHP sector. Ex standards use three groups: I (firedamp), IIA, IIB, and IIC, which correspond to the maximum experimental safe gap (MESG; equivalent to the so-called "quenching distance") or the minimum ignition current (MIC) ratio of the gas. The current classifications are from IEC 79-1: 1971, whilst earlier ones originate from BS 229 (i.e., without consideration of MIC) or variation in other countries, although different systems were sometimes used, such as based on fluid flash point only (Grover and Wilson, 1960; Grove, 1968; Riddlestone, 1968). By contrast, the development of the Globally Harmonised System of Classification and Labelling of Chemicals (GHS) began at the 1992 United Nations Rio Conference when various bodies agreed the need for a hazard classification system. This system simply identifies all flammable gases (liquified under pressure) as "Class 2.1".

An impression of the evolution of the safety classification used by RACHP sector are shown in Table 2. The earlier ASA, VBG and ASHRAE classes can be compared against the current ISO 817. Originally using three general classes ("benign", irritant and flammable), since 2014 there have been the eight classes used today. EN

378: 2000 represented a sort of transition, where the previous six classes were described, but (nonsensically) requirements were based on three groups, some of which combined higher flammability and toxicity. On one hand, this greater precision enables more specific system requirements to be formulated, on the other it increases complexity of the rules which could arguably result is errors. Whilst ISO/R 817 was first issued in 1968, it only contained refrigerant numbering; no safety classification and it was not until ISO 817: 2014 that it included safety classifications, as well as the new "2L". Similarly, safety classifications were not part of ASHRAE 34 until 1978, although the 2L subclass was added in 2010 and then changed to a class of its own in 2013. The standard also toyed with further classifications to account for fractionation of blends.

Table 2: Various refrigerant safety classifications

Group/Class	ASA B9 (1919)	VBG 20 (1959)	ASHRAE 34: 1978	ISO 817: 2014
1	Not flammable or irritant	Non-flammable without any significant toxic effect on humans	Non-flammable and no death or injury to guinea pigs during 2 h at 2.5%	No flame propagation
2L	-	-	-	Flame propagation, LFL > 3,5%, HOC < 19 MJ/kg, FS ≤ 10 cm/s
2	Irritant: any which has irritating effect on eyes, nose, throat or lungs	Toxic or corrosive and with LFL of ≥3.5%	Death or injury to guinea pigs during 2 h at 2.5%	Flame propagation, LFL > 3,5%, HOC < 19 MJ/kg
3	Flammable: any which will burn in air	LFL of < 3.5%	Flammable and no death or injury to guinea pigs during 2 h at 2.5%	Flame propagation, LFL ≤ 3,5%, HOC ≥ 19 MJ/kg
A	-	-	-	Occupational exposure limit <400 ppm
B	-	-	-	Occupational exposure limit ≥400 ppm

A fascinating article from 1915 ("Ammonia Accidents") appears to document the realisation that R717 is flammable. It refers to a meeting of the Berlin Refrigerating Society (Berliner Kaelte Verein) in 1914, where it was reported that an engineer working in an engine room "where much R717 has been lost" and "observed an advancing flame, ignited by an open light"; the engineer suffered burns to face and hands. Tests were conducted at the Technical High School in Karsruhe, which found that ignition ("with violence") occurred in concentrations between 16.5% and 26.8% with air. Three other incidents were reported in Germany (between 1909 and 1914 in Elberg, Freiburg and Berlin), although it was contested that it was actually R717 that had resulted in an explosion. One specialist argued that R717 itself wasn't flammable, but under high pressure and temperature it decomposed to hydrogen. It was concluded that pressure relief and safety valves should be used on all R717 compressors. Despite much to-ing and fro-ing, the conclusion seemed to be that R717 was in fact flammable and that regulations should be revised to prohibit open flames in engine rooms. A subsequent publication (ASRE, 1915) reported on a new regulation: "In all machinery rooms in which amounts of R717 or R160 is used, there shall be no flames, arc lights, gas jets or any apparatus employing flame". Later, Lowenstein (1916) reported "The question as to whether ammonia will burn or not has been answered by experiments." ... "When an electrically-heated platinum wire is used, an explosion results when the percentage of ammonia gas is between 19.58 and 26", validating his findings with "The author of this paper has made ammonia his life study." Nonetheless, in an article on accidents in refrigerating plants from 1917 (Starr, 1917), ten types of accidents were classified. Two of these were fire-related "ruptures caused by fires or unusual heat from outside source" and "explosions due to testing with air (oil ignition mostly)". Despite the widespread use of flammable refrigerants at the time, there were no categories related to ignition of leaked refrigerant.

In fact, it was this issue of R717 flammability that brought about the formation of the International Institute of Refrigeration (IIAR). In 1970 there was a proposed change to the USA electrical safety rules that would have mandated Ex-type equipment for R717 machinery rooms. To help oppose such requirements, the North American R717 industry initiated the IIAR.

6.2 Mechanical requirements

The evolution of RACHP standards is a little less elusive. Unlike with Ex, RACHP safety standard have remained fairly low in number, but the depth and detail (number of pages) has grown significantly. They began around 1910, with ASRE-15 (now Ashare-15[8]), followed by other countries' national standards within the following couple of decades. The first such publication in the UK was BS CP 406 British code of practice for refrigeration safety, followed by various editions of BS 4434 from 1969. The international equivalent, ISO/R 1662 – the predecessor of ISO 5149 – was not published until 1971, then as a formal ISO 5149: 1993 and subsequently in 2014. Difficulties in achieving international consensus on such a wide scope of topics is reflected in the 20+ year cycle for each revision! Table 3 contains a selection of events related to RACHP safety standards.

Table 3: Timeline and momentous events for RACHP safety rules

Year	Source	Refrigerant classification or system requirements or events
1919	ASA B9 (→ASHRAE 15)	First edition
1933	AS/NZS 1677	First published (as AS CB3-1933)
1952	BS CP 406	First edition of British code of practice for refrigeration safety
1956	VBG20 (Germany)	First issue
1957	DIN 8975	Early edition
1969	BS4434	Successor to BS CP 406
1971	ISO/R 1662	Predecessor of ISO 5149
1978	ASHRAE 34	First edition
1997	IEC 60335-2-24	Revised to included 150 g charge limit for flammables
1995	DIN 7003	Specifically covering requirements for A3 refrigerants (including HCs)
1995	BS4434 (UK)	Revised to include sensible requirements for flammables
1998	AS/NZS 1677	Published, permitting flammables
2000	SNI 6500 (Indonesia)	Published, permitting flammables
2000	EN 378	First edition
2001	NPR 7600 (Netherlands)	Safety aspects for flammable refrigerants
2002	IEC 60335-2-89	First edition, included 150 g charge limit (copy from IEC 60335-2-24)
2003	IEC 60335-2-40	Revised to include requirements for flammables
2010/13	ASHRAE 34	Inclusion of A2L classification
2014	ISO 817	Inclusion of A2L classification
2014	ISO 5149	Revised to permit flammables similar to EN 378
2018	IEC 60335-2-40	Revised to permit substantially greater quantities of A2Ls
2019	IEC 60335-2-89	Revised to include 13xLFL charge limit for flammables
2020	CDV for IEC 60335-2-40	FDIS to include better rules for flammables; to be voted on in 2022

8 A comprehensive description of the history of ASHRAE 15 can be found in Reindl (2014) and McLinden and Huber (2020) provide a description of the safety classifications from ASHRAE 15 and ASHRAE 34.

Some earlier editions of the various safety standards were less discriminatory with flammable refrigerants. For instance, early editions of DIN 8975 and VBG 20 limit flammable refrigerants to 10 kg only if occupancy exceeds one person per 8 m^2 of floor space, otherwise the limit would be 50 kg. For AC systems, they simply say that refrigerants must be prevented from entering the rooms in quantities that would be harmful. BS4434 on the other hand did not permit flammables in institutional, public or residential occupancies. Whilst ASHRAE 15 was similarly less discriminatory in the early editions, it soon moved to similarly prohibit flammables from occupied spaces. ISO 5149: 1993, was equally as restrictive, prohibiting flammable refrigerants from any situation except laboratories up to 2.5 kg and industrial "special cases".

However, with the Montreal Protocol and the re-adoption of flammables for certain sectors, various safety standards were revised, not necessarily to permit flammables, but to provide a more robust framework for their application. Thus, national standards such as BS4434, AS/NZS1677, SNI 6500, NPR 7600 and DIN 7003 (specifically for A3 refrigerants) and the first edition of EN 378 were all (re)published within a few years of each other, permitting certain quantities of flammables in all applications and occupancy types. Such requirements, including those for safety devices, safe electrical equipment, gas detection, etc., were mostly identical.

Soon after, a draft IEC 60335-2-40 was issued with extensive requirements for flammables. Various national committees objected, especially since an electrical standard (IEC 60335-) should not be so focussed on mechanical requirements, so it was agreed to have a joint working group between the responsible committee and also that for ISO 5149. The intention was to develop a consistent set of rules which could be temporarily used in IEC 60335-2-40, but then taken out and referenced in the mechanical standard ISO 5149, once it was published. (Unfortunately, this sensible intention fell afoul of the lackadaisical rate of ISO 5149 revision; by the time it was published a decade or so later, IEC 60335-2-40 had moved on further and enacting such a switch was no longer viable.) Tragically, the requirements drafted and pushed through were highly obstructive to the use of HCs and contributed to the withdrawal of otherwise safe products from the market and prevented further introduction of HC ACs for many years. A decade later, as the new 2L classification was being established, a new working group was set up to develop improved requirements based on A2L. Apart from increasing upper charge limits from about 8 kg to 80 kg and additional means of extended allow charge limits, by means of ventilation, gas detection, system tightness, etc., simplified version of Ex-type protection concepts for electrical equipment were also included. This subsequent edition was published in 2018.

The other product standard, IEC 60335-2-89, had included the 150 g limit for flammables since the first edition in 2002, despite at least one attempt to raise it a few years later. In 2014 the BSI mirror committee submitted a proposal to increase the charge limit and include additional requirements for flammables. A working group was established and eventually in 2019 the draft was submitted for final vote. Initially it seemed that the draft had failed, but once it was realised that certain countries had incorrectly submitted their votes, IEC corrected the result to approval of the standard. Whilst the recent IEC 60335-2-40, EN 378 and ISO 5149 had opened up the rules significantly for A2L refrigerants, the approval of IEC 60335-2-89 represented the first fortuity for HCs in two decades.

With the publication of EN 378, conflicting national standards had to be withdrawn. This had a nasty sting in the tail for flammables, as a single line stated they shall not be used in direct systems be permitted for "air conditioning and heating for human comfort"; the origin of this requirement is unclear. A revision of EN 378 was issued in 2008 and – despite efforts to introduce improvements – it was effectively similar to the 2000 version and the prohibition for flammables in comfort cooling was replaced with the atrocious charge limits from IEC 60335-2-40. The latest, EN 378: 2016 was not better for HCs, but did include extensive requirements for A2L refrigerants, consistent with those in IEC 60335-2-40: 2018. Again, efforts to include enhancements for HCs were unsuccessful. The slightly earlier release of ISO 5149: 2016 was similar to EN 378: 2018, both in structure and requirements for flammables. The next revision of EN 378 is expected in 2024 and will likely contain substantially improved requirements for HCs and A2Ls alike.

It is noteworthy that within the governing rules for flammables – ATEX directives or DSEAR and EPS in UK – they make no reference to limiting quantities of flammable material, just that the mitigation measures should be proportionate to the quantity and circumstances.

Until the era of re-emergence, safety standards appear to have been fairly ambivalent in their dealings with flammables. However, once it became apparent that HCs posed a threat to halogenated refrigerants' market dominance, elements of the establishment worked towards ensuring obstructive safety standards were in place, acting as a barrier (Frost, 2002; Parliament, 2004; Brown and Collins, 2002). This strategy worked phenomenally well. During the explosive era, use of flammable HFCs had to be facilitated and safety standards were revised such that they could be applied to the same extent of non-flammable refrigerants, without adverse impact on cost. However, increased interest of system manufacturers in HCs and their frustration with illogical safety standards has resulted in renewed participation in standards development, likely with an incrementally positive outcome. As is observed with general improvements in H&S as rules improved, the same must also apply to RACHP systems, including for those with flammable refrigerants. Indeed, in the earlier eras textbooks and other literature implied that whilst flammability was a concern, it was not regarded in the demonic light as it has been by some in recent times. Accordingly, the advancements in both RACHP and Ex safety standard strongly suggest systems moved from "fairly safe" to "extremely safe".

7 Reflections and outlook

7.1 Fulminating debates

The first ever refrigerant was flammable and it is likely that that most of the "last" refrigerants will also be flammable (before magnetic refrigeration or another technology takes over). Whilst nearly every book or technical articles stresses that a refrigerant "ought not", "should not" or "must not" be flammable, these refrigerants have nevertheless been, and are again, widely used on account of their thermophysical and chemical merits (good performance, material compatibility, lesser environmental impact, lower cost, etc.).

It was frequently claimed that halocarbons were commercialised in the 1930s to address the negative safety implication of the existing fluids. However, although this was evidently the case for some of the higher toxicity refrigerants, detrimental impacts of flammability are not supported by the literature (e.g., evidenced by the celebrated "propane safety refrigerant" advertisement). Furthermore, initial commercialisation of CFCs was at least partly to provide a domestic refrigerator manufacturer with a commercial (marketing) advantage over competitors. Giunta (2006) provides a comprehensive and vivid description of the events and deliberations preceding the commercialisation of halocarbon refrigerants, describing the vying of established refrigerant (R40) producers and public entities.

Due to the Montreal Protocol terminating use of CFCs and Greenpeace's Greenfreeze campaign eventually leading to the use of HCs in most of the world's domestic fridges – today, over one billion – it appears that the flammability scare-stories of yesteryear are most probably exaggerated. Whilst a notable proportion of large systems have continually used R717 since the advent of refrigeration, there has similarly been a steady growth in its use even for commercial applications.

The current renaissance of flammables, adoption and continued use of HCs, surely stems from the Greenfreeze campaign. Perversely, it is likely also to have paved the way for the need and acceptability of flammable (A2L) HFCs. Since HCs and R717 have negligible GWP, legislators saw them (and other naturals) as viable alternatives. So as regulations have edged towards low GWP, halocarbon producers were forced to find more suitable alternatives, with flammable HFCs being one of the only options on their table. Moreover, because HCs had been the "flammability punch-bag" for 25 years, it was not only easy to introduce A2Ls, but they were even offered as an antidote to high flammability HCs.

Polarisation of the topic since the Montreal Protocol was a fact. On numerous occasions, when discussing HCs, colleagues from various sectors reiterated "never", "unacceptable", "not in your wildest dreams", etc. will the industry accept flammables. Somehow vested interests managed to fire-up the emotive fear of equipment manufacturers too. In some quarters this generated hardened opposition to HCs – and more recently even to A2Ls to an extent. At times fiercely anti-HC articles were regularly found in trade literature (e.g., Anon., 2008; Anon., 2009; Anon., 2010; although far more entertaining material can be found on internet discussion forums). Presumably, reflecting on their possible hypercriticism, many commentators who were previously outspoken against HCs, have similarly opposed A2Ls. More sober sources, such as IIF-IIR (2017) should be viewed for a balanced overview of the important flammability considerations.

7.2 Candescent risks and mitigations

Experience has shown that the risk associated with the use of flammable refrigerants can be negligible and it is only with daft acts that problems arise. Of course, just as accidents involving electricity, pressurised systems and brazing/welding equipment, this is anticipated and the appropriate mitigations put in place. Improvements, such better system tightness, have undoubtedly helped, but they are incremental and not high-tech, revolutionary engineering approaches. Safely applying flammables simply requires engineers to do what they always do – design and control machinery to account for undesirable characteristics and "noise factors".

Somehow, safety standards, despite being subservient to regulations, have attained a demigod-like status, dictating "go" or "no-go" for a given fluid. Functional requirements of safety standards will likely continue to be the key to successful adoption of flammable refrigerants. Leak detection technology will most probably become a pivotal for the extended use of flammables. Mechanical gas sensors were first invented around 1800 (biological detectors being used prior to that), with the first "modern" catalytic sensors in 1926 (Halder and Chatterjee, 2019). Since then, numerous technologies have evolved and it is likely that others will further emerge.

In the current iteration of safety standards, A2Ls are given considerably more tolerance than A2s and HCs (and of course, R717 doesn't get much of a look-in anyway). For instance, maximum charge of HCs indoors is 1 kg, whereas A2Ls are permitted to reach around 100 kg (depending upon LFL). This is equivalent to a ten times larger flammable mixture, 25 times more released energy in the event of combustion and infinitely more toxic decomposition products. (Not to mention the greater likelihood and severity of explosions from adiabatic compression.) In theory, equanimity is needed, although achieving this is unlikely.

Due to the fluctuations in technology interests, RACHP safety standards have become muddled and unwieldly, with various standards spiralling off in different directions. Whilst the refrigeration system is essentially the same, substantially different rules apply depending upon the aesthetics of the housing. Unfortunately, the further these standards diverge, the more difficult it becomes to unify them.

Whilst the aforementioned discussions are almost entirely on flammability, it must not be forgotten that there are numerous other hazards associated with refrigerants that are always pertinent (Calm, 1994).

7.3 Thermogenesis

Observation of an industry "litmus paper" helps indicate the direction of things to come. Since 2015, there have been about 150 articles on the topic of safety of flammable refrigerants (as listed in IIF-IIR's FRIDOC[9]), corresponding to about 30 per year. For the preceding 20 years there were about five per year. Such a trend in research activities infers a flare up of interest. Many sectors are now accepting flammables; about 15% of the MACS sector is now using R1234yf (about 170 million cars), whilst a few million are using HCs as drop-ins. About 40 million split ACs use R32 and well over one million use R290. Dwarfing these numbers are the billion or so domestic refrigerators with R600a. Booten et al. (2020) states that HCs account for 6-7% of the global refrigerant market, R32, currently accounts for about 10% (although partly as a component in R410A) and both R717 and R1234yf about 5%. Data for Australasia/SE Asia shows sales of HC refrigerants have been doubling every 5 – 6 years. Today, some 60% of global small compressor production is for HCs, in excess of 100 million per year. Production of R1234yf compressors began around 2010, mainly for Japan, but growth for these and for other A2L blends is expected by manufacturers. Insightfully, a leading halocarbon refrigerant producer stated they expect some 30% of the refrigerant market would eventually be taken by natural refrigerants (Manzer, 1990). Considering the blends registered in ISO 817, nearly one-third are flammable as with two-thirds that have been registered over the past 10 – 15 years, most of which are primarily HFC-based.

It may be apt to make some philosophical projections about the future, based on these past eras and current conditions. Unlike for most of the time that refrigeration technology has been in existence, there presently appears to be a critical need to minimise environmental impact. Obviously the most apparent is global warming, but there are other impacts related to release of apparently indigestible compounds into the wild (Behringer et al., 2021). Previously, the refrigeration society had some degree of freedom to choose the most convenient options. But now it is increasingly evident that if we want to use refrigeration systems, it will have to be with flammables. Having said that, it is also useful to observe that there is a divergence between the drivers to adopt the different types of flammables, HCs and HFCs. The uptake in HC domestic and commercial refrigerators was (initially) driven by consumer desire for "green". With R1234yf it was legislation and with R32, manufacturer preference (reflecting a compromise between environmental impact, product cost and perceived risk). Unlike with domestic appliances in 1990s and supermarkets in the 2000s, today, refrigerant type is probably a minor criterion amongst many others for consumers of cars and ACs, considering the bombardment of current purchasing factors; refrigerant choice is lost in the fog. It is possible that for many types of equipment, refrigerant type will hold even less weight for the discerning customer. Notwithstanding, recent elevated concerns over GWP and reluctance of manufacturers to go through yet another refrigerant transition raises interest in HCs, R717 and other natural refrigerants, since they are regarded as "future proof".

My prediction is that all "smallish" systems and self-contained equipment (including big chillers) will asymptote towards flammable, negligible GWP and possibly halogen-free. For larger, distributed systems, compromise options will be used, based on the conditions (dictating temperature lift). As the industry becomes more tolerant or even accepting of flammables, they will inevitably take a larger portion. There is an approximate correlation between lower GWP, system compactness (lower cost), desirable performance and increasing flammability. Although seldom discussed, an increasing number of future options are class A2 refrigerants, so as to satisfy this compromise.

9 https://iifiir.org/en/fridoc

Soon, the "not flammable" mantra will itself become phased out, or at least phased down. Not only in the mindset of younger technicians to minimise climate impact, but familiarisation with flammables will lead to a new generation of technicians more tolerant and experienced with handling flammables. System manufacturers will become far less hesitant and even comfortable with all flammable refrigerants.

References

1. Althouse, A. D., Turnquist, C. H. 1943. *Modern Electric and Gas Refrigeration*. 4th ed
2. Althouse, A. D., Turnquist, C. H. 1944. *Modern Electric and Gas Refrigeration*. War Department Education Manual, EM977. Goodheart-Willcox Company
3. Althouse, A. D., Turnquist, C. H. 1960. *Modern Refrigeration and Air Conditioning*. Goodheart-Willcox Company
4. Althouse, A. D., Turnquist, C. H., Bracchiano, A. F. 1982. *Modern Refrigeration and Air Conditioning*. Goodheart-Willcox Company
5. Anderson, M. E. Herbert, R. H. 1971. *Refrigeration Questions and Answers*. Butterworth and Co
6. Anon, 2021. Chemours files second lawsuit over R1234yf. https://www.coolingpost.com/world-news/chemours-files-second-lawsuit-over-r1234yf/
7. Anon. 1915. *Ammonia Accidents*. Refrigeration Engineering, A. S. R. E. Journal, vol. 1
8. Anon. 1949. Schlumbohm. *Life* magazine, 4th July 1949, USA
9. Anon. 2008. *Hydrocarbon (Flammable Refrigerants – Hazardous Goods) Does NOT Belong in Cars*. VASA "Hot Air", October 2008
10. Anon. 2009. *Don't Fall for the Flammable Refrigerant Line*. VASA "Hot Air", April 2009
11. Anon. 2010. *Vehicle Repair Industry Aims to Keep Explosive Gas Out of Air Con Systems*. The Automotive Engineer, March/April, 2010
12. ASHRAE 15 2016, *Safety Standard for Refrigeration Systems*
13. ASHRAE 34 1978, *Designation and Safety Classification of Refrigerants*
14. ASHRAE 34 2010, *Designation and Safety Classification of Refrigerants*
15. ASHRAE 34 2013, *Designation and Safety Classification of Refrigerants*
16. Bäckström, M. 1946. *Kylteknikern*. Swedish Society of Refrigeration, Stockholm
17. Banks, R. E., Tatlow, J.C. 1986. *Synthesis of C-F bonds: The Pioneering Years, 1835 – 1940*. J. Fluorine Chemistry, vol. 33, Iss. 1–4, pp. 71-108
18. Behera, C., Bodwal, J., Sikary, A. K., Chauhan, M. S., Bijarnia, M. 2017. *Deaths Due to Accidental Air Conditioner Compressor Explosion: A Case Series*. J Forensic Sci. 62(1):254-257, doi: 10.1111/1556-4029.13242
19. Behringer, D., Heydel, F., Gschrey, B., Osterheld, S., Schwarz, W., Warncke, C., Feeling, F., Nödler, K., Henne, S., Reimann, S., Bleep, M., Jörß, W., Liu, R., Ludig, S., Rüdenauer, I., Gartiser, S. 2021. *Persistent Degradation Products of Halogenated Refrigerants and Blowing Agents in the Environment: Type, Environmental Concentrations, and Fate with Particular Regard to New Halogenated Substitutes with Low Global Warming Potential*. Report No. FB000452/ENG. German Environment Agency
20. Bivens, D. B. 1991. *Heat Pumps – R-22 and Beyond*. Proc. Meeting Customer Needs with Heat Pumps, report TR-101944, Electric Power Research Institute, Palo Alto, USA
21. Boast, M. F. G. 1986. *Refrigeration and Air Conditioning*. Heinmann
22. Bodio, E., Chorowski, M., Wilczek, M. 1993. *Working Parameters of Domestic Refrigerators Filled with Propane-Butane Mixture*. Int. J. Refrig., Vol. 16, No. 5, pp. 353-356
23. Boldrin, B., Minotto, G., Panozzo, G., Rebellato, L., Toniolo, B., Varotto, G. Bertocco, A. 1991. *Propane as an Alternative Refrigerant in Foodstuffs Transport*. Proc. IIF-IIR XVIIIth Int. Congr. Refrig., Montreal, Canada
24. Booten, C., Nicholson, S., Mann, M., Abdelaziz, O. 2020. *Refrigerants: Market Trends and Supply Chain Assessment*. Technical Report NREL/TP-5500-70207. Clean Energy Manuf. Analysis Center. Golden, USA
25. Breddy, N. C. 1961. *An Introduction to Refrigeration*. Draughtmen's and Allied Technician's Association
26. Brown, P., Collins, J. 2002. *Chemical Firms Promote Pollution*. Guardian, 26th July 2002. UK
27. BS 1608: 1949 *Condensing Units for Refrigeration*
28. BS 1725: 1951 *Properties of Refrigerants*
29. BS 229: 1929 *Specification. Flameproof Enclosures for Electrical Apparatus (For Use in Mines and Other Places Where an Explosive Atmosphere may be Encountered) and Tests for Flame-Proof Enclosures*

30. BS 3456: 1990 *Safety of Household and Similar Electrical Appliances. Part 202. Particular Requirements. Part 202.24 Refrigerators and Food Freezer*
31. BS 4434: 1969 *Specification for Requirements for Refrigeration Safety*
32. BS CP 406:1952 *Code of Practice. Mechanical Refrigeration*
33. Calm, J. M. 1994. *Refrigerant Safety*, ASHRAE Journal, vol. 36, no. 7, pp. 17-26
34. Calm, J. E. 2012. *Refrigerant Transitions...Again*. Proc. ASHRAE/NIST Conf., Gaithersburg, USA
35. Camporese, R., Bigolaro, G., Cortella, G., Scattolini, M. 1991. *Flammable Refrigerants in Domestic Refrigeration*. Proc. New Challenges in Refrigeration. Proc. XVIIIth Int. Congr. Refrig., Montreal, Canada
36. Carr, F. 1949. *Power Savings in Process Refrigeration*. Ind. Eng. Chemistry, vol. 41, No. 4
37. Colbourne, D., Suen, K.O., Li, T.-X., Vince, I., Vonsild, A., 2020. *General Framework for Revising Class A3 Refrigerant Charge Limits – A Discussion*. Int. J. Refrig., DOI: 10.1016/j.ijrefrig.2020.04.024
38. Collacott, R. A. 1950. *Refrigeration*. Sir Isaac Pitman and Sons
39. Crawhill, T. C., Lentaigne, B. 1934. *Refrigeration Exhibition (Guide to The)*. H. M . Stationary Office
40. Daikin, 2021. https://www.daikin.com/csr/information/influence/policy.html, last accessed November 2021
41. Davis, S., Pagliaro, J. 2019. Report No. 9013 *Experimental Study on the Consequences of Full-Scale Ignition Events Involving the A2L Refrigerant R-454C*. Air-Conditioning, Heating and Refrigeration Technology Institute, USA
42. Davy, H., *On the Fire-Damp of Coal Mines, and on Methods of Lighting the Mines so as to Prevent Its Explosion*, Phil. Trans. Royal Soc. 1-24 (1816)
43. DCLG, 2014. *Fire Statistics: Great Britain April 2012 to March 2013*. Department for Communities and Local Government. Office for National Statistics, London, UK
44. Devotta, S., Gopichand, S. 1992. *Comparative Assessment of Some Flammable Refrigerants as Alternatives to CFC12*. Proc. Int. Refrig. Air Cond. Conf., Purdue, USA
45. Dierckx, T., Berghmans, J. 1993. *Safety Aspects of Working Fluids*. Proc. IIF-IIR Conf. Energy Efficiency in Refrigeration and Global Warming Impact
46. DIN 7003: 1995 *Refrigerating Systems and Heat Pumps with Flammable Refrigerants of Group L3 – Safety Requirements*
47. DIN 8975: 1957 *Refrigeration Plants; Safety Requirements for Design, Equipment, Installation and Operation*
48. Dossat, R. J. 1961. *Principles of Refrigeration*. John Wiley and Sons, Inc.
49. Dossat, R. J. 1991. *Principles of Refrigeration*. Prentice-Hall, Inc.
50. Doyle, W. P. 2021. *William Cullen (1710-1790) School of Chemistry* http://www.chem.ed.ac.uk/about-us/history/professors/william-cullen
51. Edwards, H. D. 1922a. *Some Properties of Hydro-Carbon Refrigerants*. Proc. American Soc. Refrig. Eng., 24-26th May 1922.
52. Edwards, H. D. 1922b. *Use of Propane for Refrigeration*. Proc. American Inst. Chem. Eng. 6-9th Dec., 1922.
53. EFCTC. 2021. *European Fluorocarbon Technical Committee*. https://www.fluorocarbons.org/toxicology-and-safety/
54. Eggen, G., Lystad, T., Fagerli, B. E. 1994. *Design Criteria for Heat Pumps and Refrigerating Plants with Ammonia and Flammables as Working Fluids*. Proc. IIF-IIR Conf., CFCs, the Day After. Paduaa, Italy
55. EN 60335-2-24: 1997 *Household and Similar Electrical Appliances – Safety – Part 2-24: Particular Requirements for Refrigerating Appliances, Ice-cream Appliances and Ice Makers*
56. EN 378: 2000. *Specification for Refrigerating Systems and Heat Pumps. Safety and Environmental Requirements – Part 1 Basic Requirements, Definitions, Classification and Selection Criteria*
57. EN 378: 2008. *Refrigerating Systems and Heat Pumps – Safety and Environmental Requirements. Part 1 Basic Requirements, Definitions, Classification and Selection Criteria*
58. EN 378: 2016. *Refrigerating Systems and Heat Pumps. Safety and Environmental Requirements. Part 1 Basic Requirements, Definitions, Classification and Selection Criteria*
59. Faber, O., Kell, J. R. 1958. *Heating and Air Conditioning of Buildings*. The Architectural Press. 4th ed.
60. Fidler, J. C. (ed). 1965. *Manual of Refrigeration Practice*. Technical Publications (London) Ltd
61. Frehn, B. 1993a. *Propan als Alternatives Kältemittel für Wärmepumpen – Erste Betriebserfahrungen*. Proc. DKV-Tagungsberichte, Vol. 20 pp. 51-63
62. Frehn, B. 1993b. *Propan als Arbeitsmittel für Wärmepumpen – Die Beste Alternative zu R22*. Ki Klima-Kälte-Heizung, Vol. 10, pp 402-405
63. Frost, L. 2002. *Standards Body Faces Scrutiny Over Influence of HFC 'Cartel'*. May 29, 2002, https://www.politico.eu/article/standards-body-faces-scrutiny-over-influence-of-hfc-cartel/

64. Gandhi, P., Hunter, G., Haseman, R., Rodgers. B. 2017. Report 9007-01 *Benchmarking Risk by Whole Room Scale Leaks and Ignitions Testing of A2L Refrigerants*. Air-Conditioning, Heating and Refrigeration Technology Institute, USA

65. Gantz, C. 2015. *Refrigeration A History*. McFarland, Inc

66. Garry, M., Yoshimoto, D. 2019. 500 g. *Accelerate Magazine*, July-August 2019, Volume 1, Issue 102.

67. Gerwen, R. J. M. van, Jansen, C. M. A. 1994. *Risk assessment of Flammable Refrigerants*. Proc. IIF-IIR Int. Conf. New Applications of Natural Working Fluids in Refrigeration and Air Conditioning. Hannover, Germany

68. Giunta, C. J. 2006. *Thomas Midgley, Jr., and the Invention of Chlorofluorocarbon Refrigerants: It Ain't Necessarily So*. Bull. Hist. Chem., Vol. 31, No. 2

69. Górny, M. 2013. *History of Explosion Protection in Poland*. Stahl Ex-Magazine

70. Gosney, W. B. 1982. *Principles of Refrigeration*. Cambridge University Press

71. Grace's, 2021. https://www.gracesguide.co.uk/Main_Page

72. Granryd, E., Tengblad, N., Nowacki, J. E. 1994. *Propane as Refrigerant in a Small Heat Pump. Safety Considerations and Performance Comparisons*. Proc. IIF-IIR Int. Conf. New Applications of Natural Working Fluids in Refrigeration and Air Conditioning. Hannover, Germany

73. Grove, J. R. 1968. *The Measurement of Quenching Diameters and their Relation to the Flameproof Grouping of Gases and Vapours*. Proc. I. Chem. E. Symposium Series No. 25, London, UK

74. Grover, S. F. Wilson B. G. 1960. *The Storage and Transmission of Hazardous Liquids*. Proc. Symp. Chemical Process Hazards, Inst. Chemical Engineers, London, UK

75. Haase, H. 1977. *Electrostatic Hazards, Their Evaluation and Control*. Verlag Chemie, Weinheim

76. Halder, M., Chatterjee, S. 2019. *Microcontroller Based LPG Gas Leakage Alert System*. Int. J. Eng. and Applied Sciences (IJEAS), Vol. 6, Iss. 2, pp. 2394-3661

77. Halozan, H. 1994. *Propane – A Realistic Alternative*. Proc. IIF-IIR Int. Conf. New Applications of Natural Working Fluids in Refrigeration and Air Conditioning. Hannover, Germany

78. Haselden, G. G., Klimek, L. 1957. *An Experimental Study of the Use of Mixed Refrigerants for Nonisothermal Refrigeration*, Proc. Inst. Refrig., vol. 54, pp. 129-148

79. Hickman, K. E. 1993. *Refrigerants After CFCs – What Are the Choices*. Proc. 15th Annual Meeting Int. Inst. Ammonia Refrig. (IIAR), Vancouver, Canada

80. Higashi T., Saitoh S., Dang C., et al. 2017. *Diesel Combustion of Oil and Refrigerant Mixture During Pump-Down of Air Conditioners*. Int. J. Refrig. Vol. 75, DOI: http://dx.doi.org/10.1016/j.ijrefrig.2017.01.003

81. Higashi, T., Dang, C., Hihara, E. 2018. *Diesel Explosion of Propane at Pump-Down of Air Conditioner*. Proc. 13th IIF-IIR Gustav Lorentzen Conference, Valencia, doi: 10.18462/iir.gl.2018.1168

82. HSE, 2020. *Historical Picture Statistics in Great Britain 2020*. Health and Safety Executive, HMSO, Norwich, UK, www.hse.gov.uk/statistics/.

83. Hull, H. B. 1927. *Household Refrigeration*. Nickerson & Collins Co

84. IEC 60335-2-40: 2003 *Specification for Safety of Household and Similar Electrical Appliances. Safety. Particular Requirements for Electrical Heat Pumps Air-Conditioners, and Dehumidifiers*

85. IEC 60335-2-40: 2018 *Specification for Safety of Household and Similar Electrical Appliances. Safety. Particular Requirements for Electrical Heat Pumps Air-Conditioners, and Dehumidifiers*

86. IEC 60335-2-89:2002 *Household and Similar Electrical Appliances – Safety – Part 2-89: Particular Requirements for Commercial Refrigerating Appliances and Ice-Makers with an Incorporated or Remote Refrigerant Unit or Motor-Compressor*

87. IEC 60335-2-89:2019 *Household and Similar Electrical Appliances – Safety – Part 2-89: Particular Requirements for Commercial Refrigerating Appliances and Ice-Makers with an Incorporated or Remote Refrigerant Unit or Motor-Compressor*

88. IEC 79-1: 1971 *Construction and Verification Test of Flameproof Enclosures of Electrical Apparatus*

89. IIF-IIR. 2017. 36th *Informatory Note on Refrigeration Technologies. Flammable Refrigerants*. December 2017, Paris, http://www.iifiir.org/userfiles/file/publications/notes/NoteTech_36_EN_nkyix2fcj7.pdf

90. IOR. 2021. Guidance Note GN37: *Statement on Choice of Refrigerant*. Institute of Refrigeration, December 2021. London, UK

91. ISO R/1662: 1971 *Refrigerating Plants – Safety Requirements*

92. ISO R/817: 1968 *Organic refrigerants – Number Designation*

93. ISO 5149: 1993 *Mechanical Refrigerating Systems Used for Cooling and Heating — Safety Requirements*

94. ISO 5149: 2014 *Refrigerating Systems and Heat Pumps – Safety and Environmental Requirements – Part 1: Definitions, Classification and Selection Criteria*

95. ISO 817: 1974 *Organic Refrigerants – Number Designation*

96. ISO 817: 2014 *Refrigerants — Designation and Safety Classification*

97. James, R. W., Missenden, J. F. 1992. *The Use of Propane in Domestic Refrigerators*. Int. J. Refrig., Vol 15, No 2

98. Jones, W. P. 1985. *Air Conditioning Engineering*. Edward Arnold

99. Jordan, R. C., Priester, G. B. 1948. *Refrigeration and Air Conditioning*. Prentice-Hall, Inc.

100. Jürgensen, H. 1992. *Messungen an Haushaltskühlgeräten mit Brennbaren Kältemitteln*. Proc. DKV Jahrestagung, Germany

101. Kahlert, H. 2013. *From Good to Harm: General Motors' and DuPont's Engagement of Developing, Producing and Banning Chlorofluorocarbons*. Proc. 9th Int. Conf. for the History of Chemistry, Uppsala, Sweden

102. Kauffman, G. B. 1955. *Frederic Swarts: Pioneer in Organic Fluorine Chemistry*. J. Chem. Educ., vol. 32, no. 6, https://doi.org/10.1021/ed032p301

103. Klimek, L. 1959. *The Mixed Refrigerant Process and its Possible Industrial Applications*. Bull. Int. Inst. Refrig., vol. 5, pp. 85-95.

104. Kopylov, S. N., Kopylov, P. S., Eltyshev, I. P. 2019. *Fire Safety of 1, 2 and 2l Refrigerants: Myths and Reality*. IOP Conf. Ser.: Earth Environ. Sci., doi:10.1088/1755-1315/272/2

105. Kramer, D. 1991. *Why Not Propane?* ASHRAE Journal, Vol. 33, no. 6, pp. 52-55

106. Kruse, H., Hesse, U. 1988. *Possible Substitutes for Fully Halogenated Chlorofluorocarbons Using Fluids Already Marketed*. Int. J. Refrig. Vol 11, July

107. Kuijpers, L. J. M., de Wit, J. A., Benschop, A. A. 1991. *Experiments on the Performance of Mixtures Based on HFC152a in Domestic Freezer Equipment*. Proc. IIF-IIR XVIIIth Int. Congr. Refrig., Montreal, Canada

108. Kuijpcrs, L. J. M., de Wit, J. A., Janssen M. J. P. 1988. *Possibilities for the Replacement of CFC 12 in Domestic Equipment*. Int. J. Refrig. Vol. 11, no. 4

109. Lewis, H. C. 1943. Coyne Reference Set. *A Guide to Simplified Practical Electricity*. Coyne Electrical School

110. Lindgren, G. 2002. *Refrigerant and Method of Use in Refrigeration Systems*. United States Patent Office, Patent No.: US 6,336,333 B1. Jan. 8, 2002

111. Lowenstein, A. 1916. *Origin, Manufacture and Practical Application of Ammonia*. American Society of Brewing Technology, Vo.. 6., No. 5

112. MacIntire, H. J. 1928. S. B. *The Principles of Mechanical Refrigeration*. McGraw-Hill Book Company Ltd

113. Maclaine-cross, I. L. (1993) *Hydrocarbon Refrigerants and Motor Car Air-Conditioning*, Green Fridge Quest, School of Mechanical and Manufacturing Engineering, The University of New South Wales, Sydney

114. Maclaine-cross, I. L. (1994) *FIREBALL: A Brief Report on Pilot Experiments to Measure the Insurance Risk of Hydrocarbon Refrigerants in Motor Cars*, School of Mechanical and Manufacturing Engineering, The University of New South Wales, Sydney

115. Maclaine-cross, I. L. 2004. *Usage and Risk of Hydrocarbon Refrigerants in Motor cars for Australia and the United States*. Int. J. Refrig., vol. 27, pp. 339–345

116. Manzer, L.E. 1990. *The CFC-Ozone Issue: Progress on the Development of Alternatives to CFCs*. Science. Vol. 249, Issue 4964

117. Matthews, F. E. 1920. *The Field of Refrigerants*. Refrigeration Engineering, A.S.R.E. Journal, vol. 6

118. McLinden, M. O., Didion, D. A. 1987. *Quest for Alternatives*. ASHRAE J., col. 29, no. 12, ppp. 32-42

119. McLinden, M. O., Huber, M. L. 2020. *(R)Evolution of Refrigerants*. J. Chem. Eng. Data, vol. 65, pp. 4176–4193

120. McNally, W. D. 1943. *Safety Measures Against Lethal Gases*. Refrigeration Engineering, , A.S.R.E. Journal, vol. 45

121. Meacock, H. M. 1979. *Refrigeration Processes*. William Clovers and Sons Ltd

122. Miller, R. 1983. *Refrigeration and Air Conditioning technology*. Bennett Publishing Company.

123. Minor, B. H. 2009. *Compositions Comprising Fluoroolefin*. United States Patent. Patent No.: US 7.569,170 B2. 4th August 2009

124. Missenden, J. F., James, R. W., Wong, A. K. H. 1990. *Propane for Systems with Small Refrigerant Charge*. Proc Franco-Swedish Conf of AICVF, France

125. Moga, H. K., Parmar, A. P., Gohil, N. D. 2021. *An Autopsy Case of Death Due to AC Compressor Blast – A Rare Case Illustrating Primary, Secondary, Tertiary and Quaternary Blast Injuries*. J Forensic Leg Med. 80:102173. doi: 10.1016/j.jflm.2021.102173

126. Motz, W. H. 1932. Principles of refrigeration: *A Comprehensive Treatise on Fundamental Principles of Operation of Ice Making and Refrigerating Machinery, Properties and Values of Principal Media Used in Modern Refrigerating Apparatus*. Nickerson & Collins Co.

127. Moyer, J. A., Fittz, R. U. 1932. *Refrigeration*. McGraw-Hill Book Company Ltd
128. Nelson, C. W. 1953. *Commercial and Industrial Refrigeration*. McGraw-Hill Publishing Company Ltd
129. NGB IP Research Institute. 2014. *Patent Landscape about HFO Refrigerants and Hydrocarbon Refrigerants Used for Refrigerators and Air Conditioners* http://ipri.ngb.co.jp/news/639/
130. Nikolaevich, O. P., Ivanovich, A. L., Yurievich, M. B., Vadimovich, B. Y., Ivanovich, V. M., Nikolaevich, V. K. 1994. *RU2073058C1*, Russian Patent Office. 26th December 1994
131. Nimitz, J. and Lankford, L. 1994. *Refrigerants Containing Fluoroiodocarbons (FICs)*. International Refrigeration and Air Conditioning Conference. Paper 256. http://docs.lib.purdue.edu/iracc/256
132. Nuckolls, A. H. 1933. *The Comparative Life, Fire, and Explosion Hazards of Common Refrigerants, Miscellaneous Hazard Report Number 2375*, Underwriters Laboratories Incorporated, Chicago, USA
133. Palm, B. 2018. *Past and Future of Natural Refrigerants*. Proc. 13th IIF-IIR Gustav Lorentzen Conference on Natural Refrigerants, Valencia, Spain
134. Pannock, J., Didion, D., Radermacher, R. 1992. *Performance Evaluation of Chlorine Free Zeotropic Refrigerant Mixtures in Heat Pumps – Computer Study and Tests*. Int. Refrigeration and Air Conditioning Conference. Paper 138. http://docs.lib.purdue.edu/iracc/138
135. Parliament. 2004. https://publications.parliament.uk/pa/ld200304/ldselect/ldeucom/179/179we06.htm
136. Pearson, A. B. 2018. *The Birth of the Refrigeration Industry in London: 1850 – 1900*. Proc. Inst. Refrig., London
137. Pelto P. J., M.S. Harris, 1990. *Preliminary Assessment of the Safety Impacts of the Use of Potentially Flammable Refrigerants in Household Refrigerators*. Prepared for Appliance Research Consortium, Pacific Northwest Laboratory, Richland, W A
138. Perry, E. J. 1984. *The Presidential Address*. Proc. Inst. Refrig., 1984-85, vol. 81, pp. 12-17
139. Powell, R. L. 1991. *Long Term Replacements for R-22 and R-502: The Next Challenge*. Proc. Int. Conf. CFC and Halon Alternatives, Baltimore, USA
140. Rajadhyaksha, D., Wadia, B. J., Acharekar, A. A., Colbourne, D. 2015. *The First 100 000 HC-290 Split Air Conditioners in India*. Int. J. Refrig., Vol. 60, pp 289-296
141. Reed, G. H. 1974. *Refrigeration A Practical Manual for Apprentices*. Applied Science Publishers Ltd
142. Rees, J. 2016. *Refrigeration Nation A History of Ice, Appliances, and Enterprise in America*. John Hopkins University Press
143. *Refcom, 2021*. https://www.refcom.org.uk/news/african-explosions-send-refrigerant-warning-to-the-world/. Last accessed, November 2021
144. Reindl, D. T. 2014. *Celebrating 100 years of ASHRAE Standard 15*. ASHRAE Journal, vol. 56, no. 11, Nov. 2014
145. Rice, W., Tempe, A., Hosterman, C., Corvallis, O., Beakley, G. C. *Hydrocarbon Refrigeration System and Method*. Number: 5,056,323, United States Patent Office. 15th October 1991
146. Richards, R. G., Shankland, I. R. 1991. *Flammability of Alternate Refrigerants*. Proc. XVIIIth Int. Congr. Refrig. Montreal, Canada
147. Riddlestone, H. G. 1968. *Comparison of National Requirements for Electrical Apparatus for use in Hazardous Atmospheres*. Proc. I. Chem. E. Symposium Series No. 2, London, UK
148. Roberts, B. 2015. Daniel L Holden, Proc. CIBSE Heritage Group
149. Roger, A. 2015. *IPR Issues for Medium and Low GWP Alternatives: The Case of Intellectual Property Rights (IPR) for HCFC-22 Alternatives*. Proc. UNEP, 36th Meeting of the Open-Ended Working Group, Paris
150. SAE, 2021. *SAE CRP Alternative Refrigerants*, https://www.sae.org/standardsdev/tsb/cooperative/altrefrig.htm, last accessed November 2021
151. Sanvordenker, K. S. 1992. R-152a versus R-134a in *Domestic Refrigerator-Freezer – Energy Advantage or Energy Penalty!* Proc. Int. Refrig. Conf., Purdue University, USA
152. Schlumbohm, P. 1937. *Refrigerating System*., Serial No. 158,045 United States Patent Office, 9th August 9, 1937
153. Schlumbohm, P.1939. *New Propane Unit Provides Own Power and Refrigeration*. Refrigeration Refrigerating Engineering (ASRE Journal), vol. 27, no. 1
154. Schnitzer, H. 1983. *The Potential for Heat Pumps Working with Fluid Mixtures for the Industrial Process Heating*. Proc. IIF-IIR XVIth Int. Congr. Refrig., Paris, France
155. Seaman, H. W., Crawford, A. G. 1918. *Refrigerating, Compression Machines and Processes for. Patent No.: GB148878A*. British Patent Office, 5th February 1918
156. Sharpe, N. 1949. *Refrigerating Principles and Practices*. McGraw-Hill Book Co.
157. Shiflett, M. B.; Yokozeki, A.; and Bivens, D. B., *Refrigerant Mixtures as HCFC-22 Alternatives* (1992). Proc. International Refrigeration and Air Conditioning Conference, Purdue, USA.

158. Sicard, A. J., Baker, R. T. 2020. *Fluorocarbon Refrigerants and their Syntheses: Past to Present*. Chem. Rev., vol. 120, no. 17, pp. 9164–9303, https://doi.org/10.1021/acs.chemrev.9b00719

159. Siebel, J. E. 1918. *Compend of Mechanical Refrigeration and Engineering*. Nickerson and Collins Co.

160. Smith, C., Z. R. J. Nicholls, K. Armour, W. Collins, P. Forster, M. Meinshausen, M. D. Palmer, M., Watanabe, 2021, *The Earth's Energy Budget, Climate Feedbacks, and Climate Sensitivity Supplementary Material*. In: Climate Change 2021: The Physical Science Basis. Contribution of Working Group I to the Sixth Assessment Report of the Intergovernmental Panel on Climate Change [Masson-Delmotte, V., P. Zhai, A. Pirani, S. L. Connors, C. Péan, S. Berger, N. Caud, Y. Chen, L. Goldfarb, M. I. Gomis, M. Huang, K. Leitzell, E. Lonnoy, J.B.R. Matthews, T. K. Maycock, T. Waterfield, O. Yelekçi, R. Yu and B. Zhou (eds.)]

161. Sparks, N. R., DiLilio, C. C. 1959. *Mechanical Refrigeration*. McGraw-Hill Book Company Ltd.

162. Spatz., M. W. 1991. *Performance of Alternative Refrigerants from a System's Perspective*. Proc. Int. Conf. CFC and Halon Alternatives, Baltimore, USA

163. Stafford, E. R., Polonsky, M. J., Hartman, C. L. 1998. *Environmentalist-Business Collaboration and Strategic Bridging: An Analysis of the Greenpeace-Foron Alliance*. Proc. Seventh International Conference of Greening of Industry Network Rome

164. Starr, J. E. 1917. *Accidents in Refrigerating Plants*. Refrigeration Engineering A. S. R. E. Journal, Vol. 3, No. 5

165. Stoecker, W. F. 1998. *Industrial Refrigeration Handbook*. McGraw-Hill Education

166. Stoecker, W. F., Jones, J. W. 1982. *Refrigeration and Air Conditioning*. McGraw-Hill

167. Stonebanks, J. A. 1948. *Use of Refrigeration for Low Temperature Applications in Wartime*. Proc. Inst. Refrig., 1948-49, vol. 45, 203-220

168. Taylor, G. 2006. *Unification Achieved: William Cullen's Theory of Heat and Phlogiston as an Example of his Philosophical Chemistry*. The British Journal for the History of Science, vol. 39, pp 477-501 doi:10.1017/S0007087406008727

169. Thévenot, R. 1979. *History of Refrigeration Throughout the World*. IIF-IIR, Paris, France

170. Treadwell, D. W. 1991. *Application of Propane (R-290) to a Single Packaged Unitary Air-Conditioning Product*. Proc. Int. CFC and Halon Alternatives Conf., Baltimore, USA

171. Trott, A. R. 1989. *Refrigeration and Air Conditioning*. Butterworths

172. Tsai, Y. T., Liao, J. Y., Shu, C. M. 2018. *Explosion Characteristics of Chlorodifluoromethane and Isobutane at High Temperature and Pressure Using a 20-L Apparatus*. Int. J. Refrig. 96 (2018) 155–160

173. UBA. 2021. Umwelt Bundes Amt, Germany, https://www.umweltbundesamt.de/en/topics/climate-energy/fluorinated-greenhouse-gases-fully-halogenated-cfcs/application-domains-emission-reduction/moble-air-conditioning-in-cars-buses-railway/mobile-air-conditioning-fluorinated-refrigerants#mobile-air-conditioning-with-fluorinated-refrigerants

174. UNEP, 2012. Issue Paper: *Refrigerated Container (Reefer) Explosion*. UNEP Regional Office for Asian and Pacific (ROAP), Thailand

175. UNEP. 2015. *Report of the Refrigeration, Air Conditioning and Heat Pumps Technical Options Committee, 2014 Assessment*. United Nations Environment Programme, Nairobi, Kenya

176. UNEP. 2019. *Report of the Refrigeration, Air Conditioning and Heat Pumps Technical Options Committee, 2018 Assessment*. United Nations Environment Programme, Nairobi, Kenya

177. VBG 20, 1956. *Hauptverband der Gewerblichen Berufsgenossenschaften Sammlung der Unfallverhütungsvorschriften*, Oktober 1956. 20. Kälteanlagen

178. Wallis-Tayler, A. J. 1895. *Refrigerating and Ice-Making Machinery*. Crosby Lockwood and Co

179. White, A. G. 1922. *Limits for the Propagation of Flame at Various Temperatures in Mixtures of Ammonia with Air and Oxygen*. J. Chemical Soc. Great Britain, vols. 121 and 122, no. 719

180. Wilson, D. 1979. *The Colder, the Better*. Atheneum

181. WMO (World Meteorological Organization). 2018. *Scientific Assessment of Ozone Depletion*. Global Ozone Research and Monitoring, Project–Report No. 58, 588 pp., Geneva, Switzerland

182. Woodward, J. L. 1998. *Estimating the Flammable Mass of a Vapor Cloud*. Center for Chemical Process Safety of the American Institute of Chemical Engineers

183. Wostrel, J, F., Praetz, J. G. 1948. *Household Electric Refrigeration*. McGraw-Hill Book Company Inc.

184. Yajima, R., Taira, S., Domyo, N., Masuda, K. 1994. *The Performance Evaluation of HFC Alternative Refrigerants for HCFC22*. Proc. IIF-IIR Conf., CFCs, the Day After. Padua, Italy

185. Zgliczynski, M., Sansalvadore, P. 1994. *Contribution to Safety Aspect Discussion on Isobutane Compressors for Domestic Refrigeration*. Proc. IIF-IIR Int. Conf. New Applications of Natural Working Fluids in Refrigeration and Air Conditioning. Hannover, Germany

Acknowledgements

The author is deeply indebted to the following specialists, for their time and effort in providing useful information, documentation and memories: British Library, Els Baert (Daikin), Steven Benton (Cool Concerns), James M. Calm, John Clark (Hychill), Neil Dennis (Aecom), Rene van Gerwen (EntropyCS), Volkmar Hasse, Rolf Huehren, Kim Ihlow (DIN), Mary Koban (AHRI), Wolo Lohbeck (Greenpeace Germany), Bob Low (Koura), Jim Nelson (Refrigeration Research Museum), Björn Palm (KTH), Ari Pasek (Bandung University), Andy Pearson (Star Refrigeration), Dick Powell (Refrigerant Solutions), Douglas Reindl (University of Wisconsin-Madison), Brian Roberts (CIBSE Heritage Group), Miriam Rodway (Institute of Refrigeration), Lennart Rolfsman, Geraldine Salt (BSI), Bernhard Schrempf (KISC), William Suen (UCL), Edward Surato (New Haven Museum), Lisa Waters (Institute of Refrigeration), Marek Zgliczynski (Embraco).

About the author

Dr Daniel Colbourne FInstR

Daniel specialises in environmental, performance and safety aspects of alternative refrigerants and systems. He works on behalf of various organisations and companies, including implementing agencies, manufacturers and end users. He is a member of the UNEP Refrigeration Technical Options Committee and several BSI, CEN, ISO and IEC standardisation committees and working groups on refrigeration safety and is a member of the Institute of Refrigeration Technical Committee.

Appendix Table 4: Characteristics of flammable refrigerants

Refrigerant	Name	Chemical formula	"Discovered"	Mol mass (kg/kmol)	NBP (°C)	Safety class	LFL (% vol.)	HOC (MJ/kg)	AIT (°C)	MIE (mJ)	FS (cm/s)	ODP (R12=1)	GWP (100 y)
R30	Methylene chloride	CH_2Cl_2	1839	84.9	39.6	B2	14.1	5.7	662	10000	n/k	n/k	11.2
R32	Methylene fluoride	CH_2F_2	c. 1930	52.0	-51.7	A2L	14.4	9.5	645	n/k	6.7	0	771
R40	Methyl chloride	CH_3Cl	1835	50.5	-24.2	B2	10.7	12.8	632	n/k	n/k	0.015	5.5
R41	Methyl fluoride	CH_3F	1835	34.0	-79.0	(A2)	7.1	19.6	n/k	n/k	28	0	135
R50	Methane	CH_4	1777	16.0	-161.5	A3	5.0	50.0	645	0.29	40	0	27.9
R141b	1,1-Dichloro-1-fluoroethane	CH_3CCl_2F	c. 1930	117.0	32.1	(B2)	7.6	8.0	532	n/k	n/k	0.10	860
R142b	1-Chloro-1,1-difluoroethane	$C_2H_3ClF_2$	c. 1930	100.5	-10.0	A2	8.0	8.9	750	n/k	4.6	0.065	2300
R143a	1,1,1-Trifluoroethane	CH_3CF_3	1948	84.0	-47.2	A2L	8.2	10.3	750	n/k	7.1	0	5810
R152a	1,1-Difluoroethane	CH_3CHF_2	c. 1900	66.0	-24.0	A2	4.8	16.3	455	n/k	23	0	164
R160	Ethyl chloride	C_2H_5Cl	1759	64.5	11.8	(B3)	3.7	19.9	507	2.37	n/k	n/k	0.5
R161	Ethyl fluoride	CH_3CH_2F	c. 1900	48.1	-37.6	(A3)	3.9	n/k	n/k	n/k	n/k	0	5
R170	Ethane	C_2H_6	1834	30.0	-88.6	A3	3.1	47.5	515	0.26	47	0	0.4
RE170	Dimethyl ether	CH_3OCH_3	1275	46.1	-24.4	A3	3.0	28.8	235	0.32	54	0	<1
RC270	Cyclopropane	C_3H_6	n/k	42.1	-32.9	(A3)	2.4	20.9	498	0.17	52	0	<1
R290	Propane	C_3H_8	1857	44.1	-42.1	A3	2.1	46.3	470	0.25	45	0	0.02
R600	Butane	C_4H_{10}	1849	58.1	-0.5	A3	1.6	45.7	365	0.25	45	0	0.006
R600a	Isobutane	C_4H_{10}	1866	58.1	-11.7	A3	1.8	45.6	460	0.25	37	0	0.006
R601	Pentane	C_5H_{12}	n/k	72.2	36.1	A3	1.45	44.8	264	0.22	43	0	<1
R601a	Isopentane	C_5H_{12}	n/k	72.2	27.9	A3	1.4	44.9	420	0.21	36.6	0	<1
R610	Ethyl ether	C_4H_{100}	<1540	74.1	34.9	(A3)	1.9	33.8	180	0.20	33.8	0	0.01
R611	Methyl formate	$C_2H_4O_2$	1671	60.1	31.8	B2	5.0	15.3	459	0.4	0.5	0	"low"
R717	Ammonia	NH_3	<1774	17.0	-33.0	B2L	16.7	18.6	630	680	7.2	0	0
R1270	Propene	C_3H_6	c. 1850	42.1	-47.6	A3	2.7	45.8	457	0.26	48	0	0.02
R1150	Ethylene	C_2H_4	1795	30.0	-103.8	A3	2.7	47.2	470	0.10	71	0	<1
R1130(E)	Trans-1,2-dichloroethene	$C_2H_2Cl_2$	< 1933	96.9	47.6	B2	5.6	10.0	460	n/k	n/k	0	"low"
R1132a	1,1-Difluoroethylene	CF_2CH_2	< 1943	64.0	-83.0	A2	5.0	15.6	n/k	n/k	n/k	0	0.05
R1234yf	2,3,3,3-Tetrafluoropropene	$C_3H_2F_4$	1945	114.0	-29.4	A2L	6.2	10.7	n/k	n/k	1.5	0	0.05

"n/k" means not known, "discovered" or first synthesised dates are very approximate, safety classifications in parentheses () are assumed by the author, references (in order of priority): ISO 817: 2014; EN 378: 2016; Smith et al., 2021; WMO, 2018; UNEP, 2019; Woodward, 1998; Haase, 1977; Internet, 2021

Nomenclature. NBP – normal boiling point (at standard atmospheric pressure) (°C), Safety class – refrigerant safety classification according to ISO 817, LFL – lower flammability limit (% vol.), HOC – heat of combustion (MJ/kg), AIT – Auto-ignition temperature (°C), MIE – Minimum ignition energy (mJ), FS – laminar flame speed (cm/s), ODP – Ozone depletion potential, with the reference R12=1 (-), GWP – global warming potential based on 100-year time horizon ($kgCO_2$-eq)

DISCUSSION REPORT

Dermot Cotter, chairing, thanked the author and commented, that he liked the way the author had split down your eras into four. He asked the author what kind of quantities of refrigerant do you believe they had in their systems? Was it small quantities or large quantities? Was there anything in the history that you could gather on that?

Daniel Colbourne
That's a very good question. Even though most of my life seems to revolve around dealing with refrigerant charge amounts, that isn't something that I managed to uncover. I'll save that investigation for part two. I guess that it ranged from relatively small quantities to tens or hundreds of kilos. Looking at some of the photographs of very old equipment, there were huge, several metre diameter flywheels and so forth, surely, they must have had very large refrigerant charge sizes.

Dermot Cotter
Where these were originally put in, they would have been in the very wealthiest people's houses and therefore they would have had big rooms and less ignition sources probably than today in a lot of ways.

Daniel Colbourne
A lot of houses 100 years ago didn't have electricity, they just had gas. In our house there are still fittings for gas lights coming out of the walls, so maybe there were a lot of ignition sources.
But it's a good question. I think more homework is needed to look at the charge amounts.

Dermot Cotter
And then you moved from that era to where flammables nearly disappeared except for ammonia. I was reading it as you were indicating that was there something stopping them coming back in. Was there lots of barriers? Was those selling the HFCs not wanting them to go back into the marketplace? Was there a little bit of challenge?

Daniel Colbourne
You mean before the Montreal Protocol?

Dermot Cotter
Yes.

Daniel Colbourne
The more you dig the more information you uncover, obviously. For example, there was some indication that the commercialisation of CFCs was initially to give one domestic fridge manufacturer, Frigidaire, a marketing advantage over their competitors. But actually, the company that commercialised CFCs was called Kinetic Chemicals.
There's a study in 1933 on behalf of Kinetic Chemicals, where they actually carry out very detailed flammability and toxicity studies on all the competing refrigerants that were used in household refrigeration. Later, I guess people just got used to using non-flammable, lower-toxicity refrigerants. Why would you make life more difficult than it already is? So, I suspect that it's just what people were used to really.

Dermot Cotter
So now there is a 150g limit out there. If you were able to, what would you change the limit to yourself?

Daniel Colbourne
In general, I don't think there should be upper charge limits. Allowable charge should be determined by risk assessments and, in fact, the limit of 150g potentially cause a huge barrier to the industry because it has taken on some sort of universal significance – now 150 g is "go or no go" for using a refrigerant which is absolutely daft. It reminds me of Xeno's bald man paradox - You've got a man who's got a full head of hair and you start picking hairs out, one-by-one. But at what point is he a bald man? And I think it's the same with charge size limits; why is 149g "safe" and 151g "not safe"? Most practitioners know about risk assessment; you work out the best, the most appropriate charge for the system. You can carry out tests and experiments to determine what quantity provides sufficiently low risk for your particular application.

Dermot Cotter
So, in one way, for a product that you produce, like a fridge or domestic freezer or a supermarket cabinet, it's quite easy to do that risk assessment because they employ experts like yourself. When it comes to equipment like a split air conditioning units where you have people who are installing it but they don't necessarily read EN378 because it's too expensive, and they don't read DSEAR - that is where the risks are. How would you deal with that situation better or would you want qualifications and examinations for putting in flammable refrigerants over a certain size of system?

Daniel Colbourne
Obviously, you get markings and warnings on equipment. But I think if you take air conditioning, it's almost self-regulating. In my little office here, I wouldn't install a 30kW split air conditioner with 3 kilograms of refrigerant. I'd maybe have a 1kW air conditioner with 130 g of refrigerant that would pose an extremely small risk in the event if a leak. Of course, you have some situations where, for example in some hotels, you have big ice makers that are placed in a little compartment room, where such a situation of a large refrigerant quantity relative to the room size could arise, but the manufacturers of equipment know where their equipment is installed and they make the appropriate design decisions associated with the anticipated use.

Dermot Cotter
Would you be happy putting flammables into vehicles instead of A2Ls in the air conditioning? Would you be happier to put a hydrocarbon into a vehicle rather than the present A2Ls?

Daniel Colbourne
The number of cars that have been retrofitted with, or not even retrofitted, they just have HCs dropped into them is thought to be significant (tens of millions), and yet there have been no safety issues reported. So, I don't think using HCs in car air conditioning systems that have been designed for them is unsafe.

Dermot Cotter
So, when it comes to flammables, what do you think is the future for heat pumps? Is it R290? Or what other refrigerants do you think from the flammable range would you be encouraging?

Daniel Colbourne
If you look at some sectors where the consumer has had the choice, or some buying power, there seems to be a preference to go for the more environmentally friendly option. So, with heat pumps, we can see, in fact, in the 1990s and very early 2000s, there was an awful lot of R290 heat pumps. It then sort of flipped back to HFCs due to problems arising from the introduction of the Pressure Equipment Directive preventing the use of non-approved compressors, but now there are something like 170 domestic heat pump products available, using R290. The consumer wants to use a heat pump because there are environmental advantages. Why not use a natural refrigerant in that? I

think that the manufacturers' output reflects that there are some types of systems where that is more challenging but it's an engineering problem and I think that the industry will find a way around it. I'm pretty sure most products will go towards A3s. Our heat pump is R290, I'm very pleased to say.

Dermot Cotter

Someone listening online has stated that the first R32 split air conditioner that they installed was in 2009. There's always going to be something that you don't know about, but it's good new information so if you ever write a book on the subject, it would be good if that person with that knowledge would please let Daniel know about it because I think it's useful information.
What about, in the global markets, is there more service technician training required? Is that really the challenge actually prohibiting the use of flammables, is it actually the skills of technicians?

Daniel Colbourne

Of course, not all technicians are competent to handle all flammables. I can't remember the numbers, I'm sure some in the e audience have got a much better idea of the proportion of technicians that have been trained or certified to handle flammables. But, there'll be a sort of generational shift. As more and more younger technicians enter the industry they'll already be trained and competent in handling flammables and more of the equipment that they work on will contain flammables and over the years it will become business as usual like it was with non-flammables before.

Dermot Cotter

Another question that's come in is asking you to comment on odorants, we understand that it's a struggle for most refrigerants.

Daniel Colbourne

Yes, it's a very interesting subject and, in fact, a number of years ago, we tried to start an industry project on odorant for hydrocarbons. We didn't get the funding, but it's still something to bear in mind. Interestingly, one of the chemical producers, developed an odorant for refrigerants in the 1990s or late 1980s but I was told they decided to shelve it because they realized that it would inhibit their refrigerant sales, which, of course, it would. We also tried looking at something similar to what is used in LPG and natural gas today. In the refrigeration system it tends to be depleted over a period of time and we found that, after a few months, you can't smell it. It's either being degraded due to the compressor discharge conditions and/or is being absorbed into the oil.
In Australia, in fact, they have a law that says that the flammable refrigerant has to be odorised. American AHRI recently published a report looking at possible odorants for flammables. I think it was a little bit inconclusive but there are a number of candidates out there and I think it is an interesting line of investigation. Technician interaction with the system is the greatest area of risk when handling flammable refrigerants. Obviously, it wouldn't address the compressor explosion issue that I mentioned earlier, but as far as releasing of refrigerant, I think it would be a major benefit to extending their use.

Dermot Cotter

And your final question - The EU Green Deal, will it accelerate or delay the future use of flammables? My belief is, there is a massive future for the refrigerants, and this could encourage their use even more. That's my opinion.

Daniel, it has been an excellent presentation. Thank you very much.

IOR.org.uk
REFRIGERATION AIR CONDITIONING HEAT PUMPS
The knowledge hub for refrigeration, air conditioning and heat pumps

Improving Cooling and Oil Distribution in an Oil-Injected Screw Compressor

Nausheen Basha Ph.D.

Winner of the IOR Ted Perry Research Award 2021

Why you should attend

- Hear about the IOR Ted Perry Award Winning student's research into performance and reliability of oil-injected screw compressors.
- Learn how the oil injected in a twin screw compressor distributes itself in the compression chambers.
- Find out how CFD (computational fluid dynamics) were used as part of this research.
- Explore the results of case studies, tests on existing rigs and instrumental detail which show the effect of using different injection points and how to reduce power losses due to temperature rise.

Abstract

Air compressors are widely used in the manufacturing, process, construction, and energy industries. They consume nearly 20% of the electricity generated worldwide [1]. The distribution of oil injected in an air compression chamber is critical for performance and reliability. Accordingly, investigations are carried out using Computational Fluid Dynamics (CFD), based on a Volume of Fluid (VOF) model, to determine the oil distribution in an air compressor. The predictions were compared with test results. This was applied to an industrial air compressor with a traditional single oil injection point. Using different nozzle diameters allowed the evaluation of the oil distribution and temperature close to rotor surfaces to be determined. It was found out that the high gas temperatures coincided with the regions of low oil concentration. With increasing the oil flow rate through a single port, the cooling of these hot spots was not achieved beyond a certain point unless the second oil-injection port was introduced on the opposite rotor. The injection through ports on both rotors reduced the compressor chamber temperature by 30° – 35°C and the specific power by 1.8%. This study shows that computational methods could be exploited for improving compressor energy consumption through oil distribution.

Introduction

Positive displacement rotary machines can be classified into various types based on the number of shafts. The ones with the single shaft are sliding vane, liquid ring and scroll machines. The others with two shafts are roots and twin screw machines and three shafts are single screw machines. The focus of this research is twin-screw air compressors, these can be used for refrigeration and air-conditioning systems, provided the working fluid is a refrigerant. These types of rotary positive displacement machines are preferred for applications where low discharge pressure and a broad range of air mass flow rates are needed. In the UK alone, compressed air systems consume 10% of industrial energy, releasing 470 kt CO_2 each year [2]. This significant amount of CO_2 emissions coming from air compressors has irreversibly damaged our environment. Thus, a way forward to reduce emissions is by reducing compressor power consumption or improving its efficiency.

Within twin-screw air compressors, the most common are oil-injected compressors with an 88% share. The large share is due to these machines achieving high pressure within a single stage of operation. This high pressure in the compression chamber results in a temperature rise. For instance, an oil-free twin-screw compressor with working fluid as air and operating at a speed of 6000 rpm and pressure ratio of 6, will result in an average discharge temperature higher than 320°C[3]. Oil injection can substantially reduce this temperature. Thus, the oil serves the important purpose of cooling the compression chamber. Also, the oil seals leakage gaps improving volumetric efficiency and lastly, oil lubricates the rotors and the bearings. The rule of thumb in the screw compressor industry is that oil contribution by mass to cooling, sealing and lubrication is in the proportion of 100:10:1, respectively. Although injecting oil has many advantages, too large or too low oil injection quantities can result in additional power losses [4]. Therefore, optimising oil injection parameters is needed for improving compressor performance.

Analytical studies carried out indicated that oil droplet size had an effect [5], but oil mass flow rate had a larger effect [6]. These studies were based on the assumption of overall heat transfer rates between the air and oil, and are insensitive to reflect on the performance change when the oil-injection port position or the number of ports are varied while maintaining the same oil injection mass flow rate [6][7][8]. Additionally, the experimental studies studied the effect of oil to gas mass ratio at varying rotational speeds [9], and also investigated how this varied both with shaft speed and discharge pressure[10]. These showed that an optimum quantity of oil exists for a compressor running at a specific operating condition but did not account for how it is distributed within the compression chamber. This only became possible through the development of Computational Fluid Dynamics (CFD) analysis of flow within these machines.

CFD analysis of twin-screw machines was made possible due to the methodology developed to produce block-structured grids for deforming domain was derived for twin-screw machines [11] and this was further developed to obtain a single domain mesh [12]. Then, Rane et al., analysed the flow in an oil-injected compressor through an Eulerian-Eulerian multiphase approach, which allowed visualising the distribution of oil for the first time within the compression chamber [3]. Various multiphase models could be used to resolve air-oil two-phase flows within the compression chamber. A typical oil-injected air compressor was investigated to compare various multiphase models [13]. It was shown that both power and flow rate predictions from the Volume of Fluids (VOF) model were close to experimental data, and solutions with the VOF model required 25% less time than the full Eulerian-Eulerian model [14]. Therefore, the VOF model was selected for further, more detailed studies.

In the study, oil and temperature distribution are explored in the chamber through CFD and VOF modelling. The performance of the compressor is evaluated when oil is admitted through the single port at various oil mass flow rates and compared with the experimental data. Based on this, the improvements in oil injection port design is made for reducing high-temperature spots in the compression chamber.

Computational study

Computational domain and mesh

The rotors of the oil-injected compressor are with 'N' rotor profile with a 4/5 lobe combination. The male rotor outer diameter is 98.8 mm, the female rotor outer diameter is 77.8 mm, the built-in volume ratio is 4.6, and the oil port position with male rotor angle is 68.7°. The compressor geometry is simplified by reducing the mechanical elements like bearings, seals, plugs etc. The compressor elements included in the model for CFD analysis include the male rotor, female rotor, clearances, suction port, discharge port and oil injection ports. The extracted internal fluid domain is shown in Figure 1a. The nominal interlobe, radial and axial leakage gaps are 50 μm. The oil injection port was located only on the female rotor side of the casing.

The screw compressor domain comprises the deforming and stationary fluid domains, as shown in Figure 1c. Grids for the moving fluid domains around rotors were generated in several transverse planes using the in-house software SCORG. The 3D mesh is generated by assembling 2D grids from these transverse cross-sections. One of the transverse 2D grids is shown in Figure 1b. The mesh movement is achieved through the process of node-mapping [15].

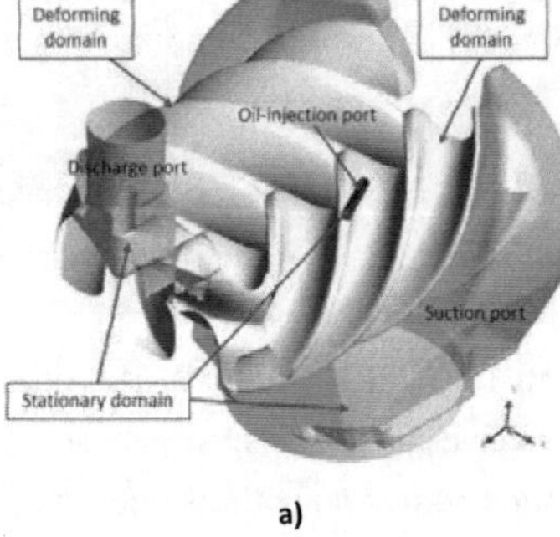

a)

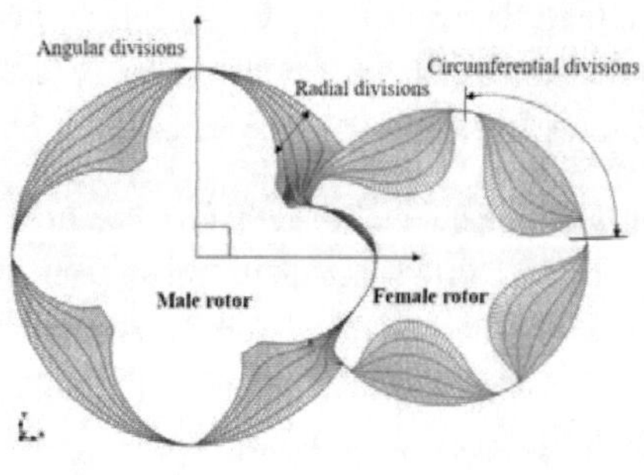

b)

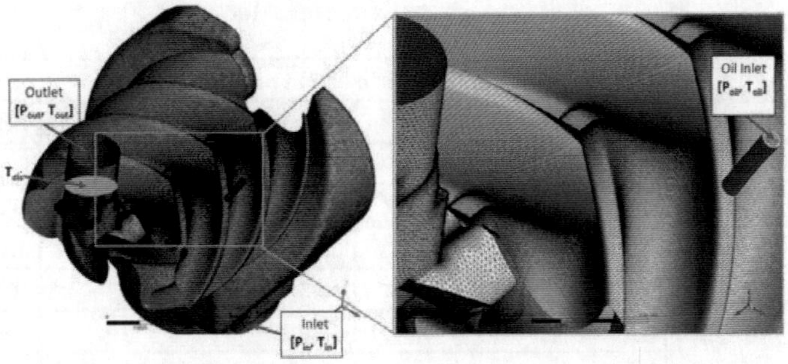

c)

Figure 1: (a) Computational domain of the compressor, (b) Rotor mesh in the transverse plane at 0° male rotor rotation and (c) Grid structure of the screw compressor

The rotor grids for this test case, which are hexahedral cells, were generated with 60 circumferential divisions, 5 radial divisions, 90 angular divisions and 77 interlobe divisions that resulted in 924120 elements. Numerical meshes for the stationary domains are tetrahedral cells generated from Ansys Mesher.

Solver setup and test cases

The air medium in the compression chamber is an ideal gas and oil is injected with the constant density of 950 kg/m^3. The specific heat capacity, viscosity, and thermal conductivity of air is 1004.4 J/kg/K, 1.831x10^{-5} kg/ms and 2.61x10^{-2} W/Mk, and oil is, 1800 J/kg/K, 0.08 kg/ms, 0.18 W/Mk, respectively. The inlet pressure is set at 0.0 bar and outlet pressure at 8.5 bar/10.5 bar depending on the operating condition. Oil is injected according to the outlet pressure of 8.5 bar/10.5 bar and injection temperature of 50°C.

y+ in the domain was in the range of 0.5 – 62.0, and for this wide range of y+, a hybrid turbulence model like SST k- ω was suitable. The transient analysis was carried out with the constant time step of 2.778 x10^{-05} s. The formulation for modelling two-phase is the VOF model and more details can be found in the manual [16] and a paper published by the author details setting for numerical parameters, relaxation factors and convergence criterias [17]. This VOF model assumes that the air and oil are immiscible in a typical oil-injected air compressor. However, suitable assumptions in the model need to be considered if the working had been a refrigerant due to its high solubility and oil affecting the fluid viscosity.

The simulation was run for 0.051-0.083s, which reflected in 20-33 full rotations of the male rotor.

Table 1: Simulated case studies

Single oil injection port			
doil [mm]	**$\dot{m}$oil /$\dot{m}$air**	**Pout [bar]**	**Poil [bar]**
3	1.6	8.5	8.5
	1.8	10.5	10.5
4	3.2	8.5	8.5
	3.8	10.5	10.5
5	5.3	8.5	8.5
	6.3	10.5	10.5
8	15.3	8.5	8.5
	17.4	10.5	10.5
Two oil injection ports			
doil1= 3.5 and doil2= 3.6, equivalent doil= 5	5.3	8.5	8.5
	5.7	10.5	10.5

The overview of case studies for various injection port diameters is shown in Table 1. All case studies were calculated for 6000 rpm. The mass flow rate of air and oil shown in the table is calculated as the averaged mass flow rate at the inlet and outlet.

Experimental investigation

The screw compressor used in this study is measured in the test rig installed at City, University of London to determine the air flow rate, compressor power, specific power, adiabatic and volumetric efficiency. The photograph of the test rig and the instrumentation diagram is shown in Figure 2. P, T, T_{ql}, and S are respectively the pressure, temperature, torque and speed measurement points in the compressor test rig. Measurements are performed according to ISO 1217:2009 [18]. The air flow rate was measured using an orifice plate, according to BS 1042 [19]. More details related to the rig can be found in the reference [20]

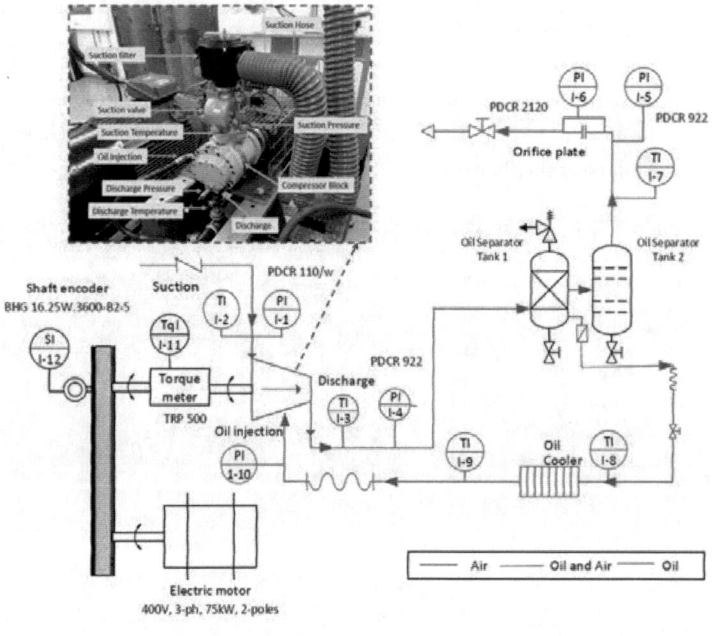

Figure 2: Layout and instrumentation of the compressor test rig

The compressor inlet and outlet temperatures were measured by Platinum Resistance Thermometers with errors within ±0.50 °C. Oil temperatures were measured by K-type thermocouples with errors within ± 1.0 °C. All pressures were measured with transducers with errors within ± 0.6%. The compressor speed was measured by a shaft encoder with ±2.7% error. The compressor torque was measured by the torque meter with a strain gauge transducer with an accuracy of ±0.25%. All tests were carried out at the rotational speed of 6000 rpm. The indicated power is 85% of the total power as it was assumed that mechanical losses are 15%. The error analysis was performed according to [21] and [20]. It resulted in the following accuracy: air flow rate 3.7%, power input, 2.7%, specific power 5.5%, and oil flow 12.8%.

CFD results and discussions

Oil and temperature distribution for a single-point injection

The oil distribution shown by the oil volume fraction and the gas temperature in the first layer of numerical fluid cells on the rotor surface is shown in Figure 3 and Figure 4, respectively.

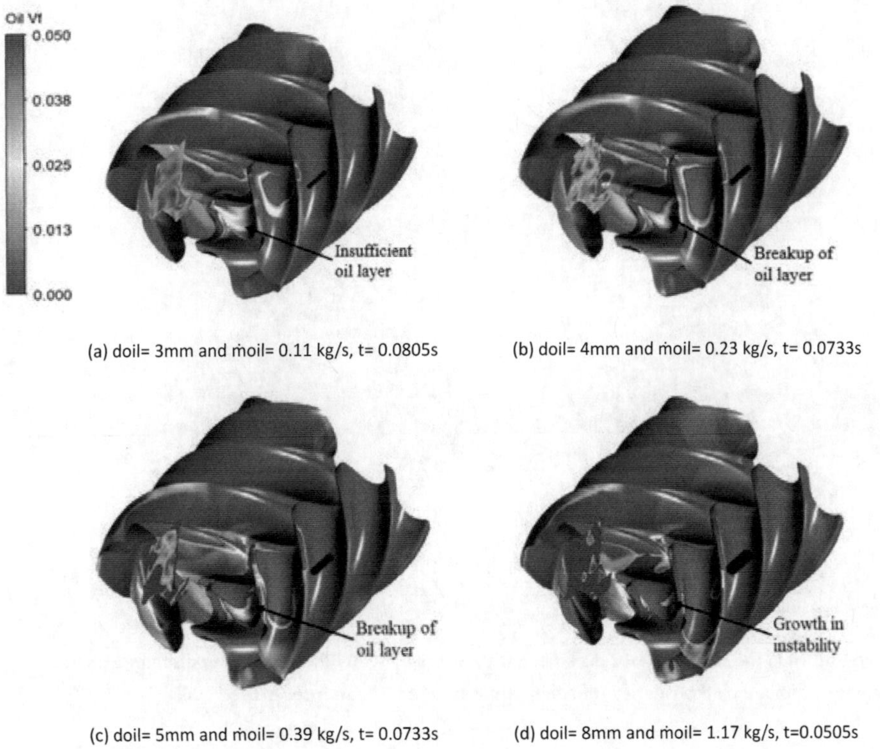

(a) d_{oil}= 3mm and $\dot{m}_{oil}$= 0.11 kg/s, t= 0.0805s

(b) d_{oil}= 4mm and $\dot{m}_{oil}$= 0.23 kg/s, t= 0.0733s

(c) d_{oil}= 5mm and $\dot{m}_{oil}$= 0.39 kg/s, t= 0.0733s

(d) d_{oil}= 8mm and $\dot{m}_{oil}$= 1.17 kg/s, t=0.0505s

Figure 3: Oil volume fraction in the first layer of cells adjacent to rotors at 6000 RPM and 8.5bar discharge pressure

The 3mm nozzle does not supply sufficient oil and the gas temperature locally rises to 220°C (Figure 3a and Figure 4a). At the same time, the discharge temperature exceeds 110°C. For the 4mm oil injection hole, the local gas temperature on the female rotor leading tip exceeds 100°C (Figure 3b and Figure 4b). While for the 5mm oil injection, the female rotor side is better cooled (Figure 3c and Figure 4c). However, in both cases, the male rotor side experiences hot zones above 100°C with temperatures in some regions as high as 189°C for 4mm nozzle and around 157°C for 5mm nozzle, though the discharge temperature is below 90° for both the cases.

Changing the orifice from 5mm to 8mm increases the oil flow by a factor of 2.8 but results in a discharge gas temperature reduction of only 11°C. The excess of oil does not contribute sufficiently to the cooling of the gas. An important point to note is that, even with this highest oil mass flow rate of 1.17kg/s, the distribution of oil on the male rotor side remains sparse, and the temperature still exceeds 100°C in certain regions (Figure 3d and Figure 4d). This observation of high temperatures in the compression chamber is also similar to the outlet pressure of 10.5 bar [17].

It can be concluded that, with the single oil injection port on the female rotor side, the oil layer sufficiently spreads on the female rotor side, but the oil does not spread effectively on the male rotor. Hence, the increase in the oil flow rate helps to reduce the gas discharge temperature, which drops down to 50°C, but the local temperature inside the domain on the male rotor side remains above 100°C.

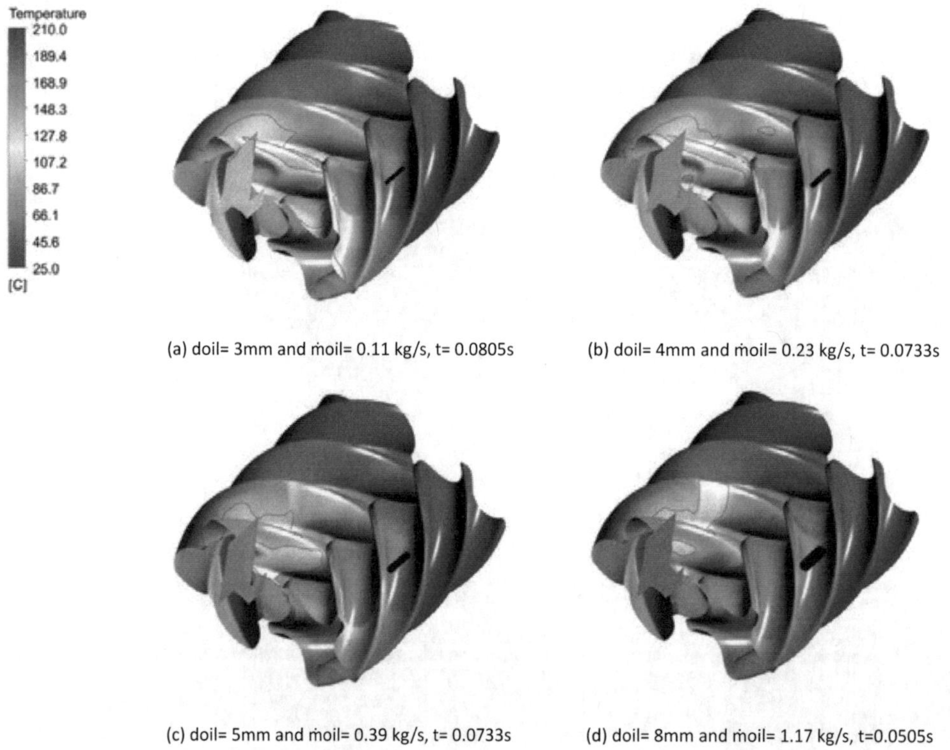

(a) doil= 3mm and $\dot{m}$oil= 0.11 kg/s, t= 0.0805s (b) doil= 4mm and $\dot{m}$oil= 0.23 kg/s, t= 0.0733s

(c) doil= 5mm and $\dot{m}$oil= 0.39 kg/s, t= 0.0733s (d) doil= 8mm and $\dot{m}$oil= 1.17 kg/s, t=0.0505s

Figure 4: Local gas temperature in the first layer of cells adjacent to rotors at 6000 RPM and 8.5bar discharge pressure. The regions bounded by the isolines represent areas where temperatures exceed 100°C

To appreciate the significance of the surface temperatures close to rotors, the total surface areas with temperatures higher than 100°C, are summed up for various oil injection diameters in Figure 5. A large difference between the surface areas with high temperatures on the male and female rotors can be observed, thus emphasising the need for better distribution of the oil to transfer the heat more evenly.

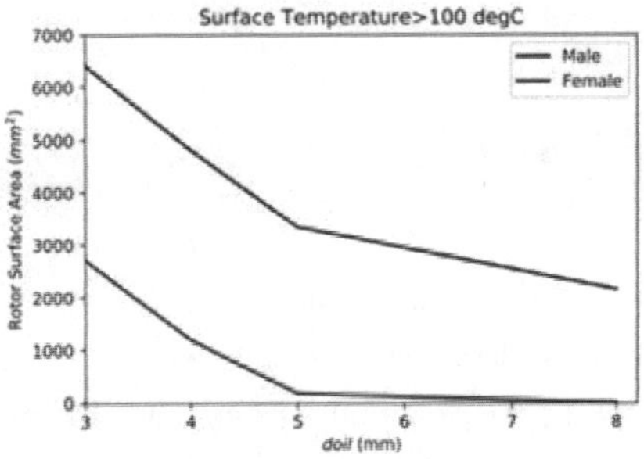

Figure 5: Distribution of temperatures of air>100°C on the male and female rotor surface area shown with the diameter of the oil injection port

Effect on the compressor performance for a single-point injection

The comparison between the performance predicted from CFD calculations and the performance measured in the test rig is shown for 5mm oil injection hole (Figure 6). The difference for the air flow rate is 0.5% at 8.5bar and 0.7% at 10.5bar, while the difference for power is around 3.0% at 8.5bar and 2.3% at 10.5bar. This shows that the CFD results are comparable to the experimental data.

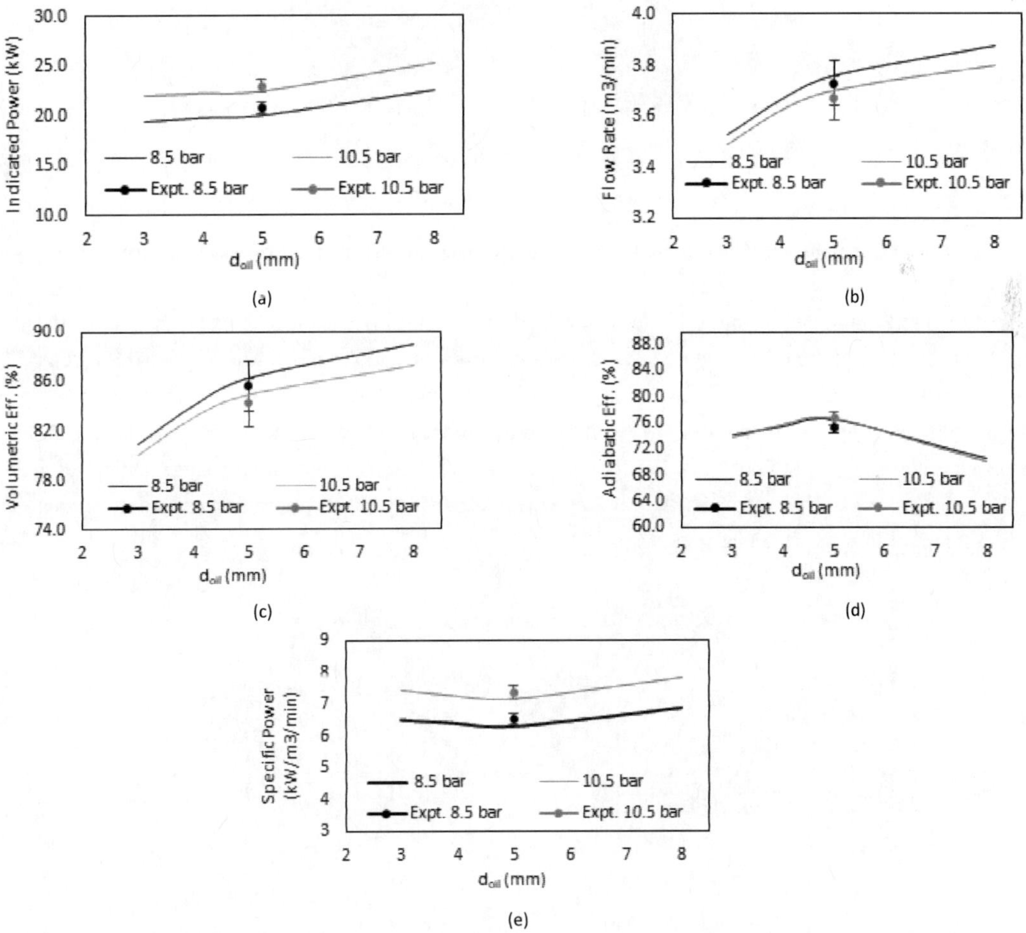

Figure 6. (a) indicated power, (b) air flow rate, (c) volumetric efficiency, (d) adiabatic efficiency and (e) specific power and with oil port diameter for discharge pressure of 8.5bar and 10.5bar

Based on the above observations, a separate injection port on the female side of the compressor is explored with the aim of improving distribution in the next section.

Oil injection through two injection ports

An additional oil injection port was added in the casing on the male rotor side to reduce the high internal gas temperatures with better oil distribution. The cumulative oil flow rate from the two injection ports is made equal to the optimum oil flow rate calculated for the single injection port, and the port is positioned to match the injection angle similar to the female rotor. Oil distribution and temperatures for two-point oil injection.

The CFD calculation results for the single injection port and two injection ports are presented in Figure 7, showing the oil distribution, and in Figure 8, the temperature distribution. Both are at the discharge pressure of 8.5bar and the male rotor rotational angle of 30°.

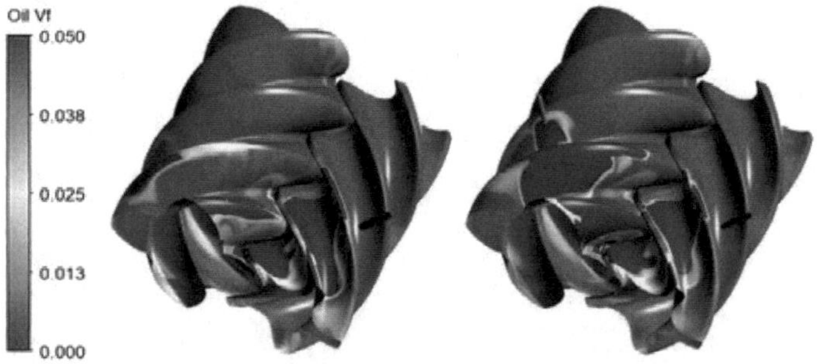

Single Injection Port   Two injection Ports

Pout= 8.5bar at t= 0.0733s

Figure 7: Comparison of oil distribution for the single injection port and the two-injection port

Looking at the left Figure 7, when the oil is injected through the single port, most of the oil flows towards the outlet. However, some of the oil is directed towards the male rotor tips due to the inertial forces. This results in sparse oil distribution on the male rotor, and high temperatures of air between the male rotor lobes. For the single-point injection, the temperature between the male rotor lobes exceeds 100°C reaching up to 157°C. However, when oil is injected through two injection ports, the distribution of oil in the working chamber of the male rotor is improving as the area in which the temperature exceeds 100°C is significantly reduced, and the maximum temperature is reduced from 157°C to 127°C.

Single Injection Port   Two injection Ports

Pout= 8.5bar at t= 0.0733s

Figure 8: Comparison of the gas temperatures for the single injection port and the two injection port

The separate injection to each rotor clearly reduces the wetted female rotor region area and supplements an increase in the wetted area near the male rotor region. It can be seen from Figure 8 that the use of two injection ports drastically alters the local temperatures. Both are at the discharge pressure of 8.5bar. A similar observation is seen for the discharge pressure of 10.5bar when the injection through two-points reduces the maximum temperature from 150°C to 115°C compared to the sing injection point.

Effect on the compressor performance for two injection points

Performance improvements due to the use of two oil injection ports are shown in Figure 9. The addition of the second port did not influence the air flow rate. At the same time, the oil flow rate was reduced by 9%. The main effect on the performance was by reduction in power consumption for the two injection ports. Indicated power reduced by 2.1% and 1.8% for 8.5bar and 10.5bar discharge pressure, respectively. This combined effect of flow and power reduced specific power by around 1.8% for both 8.5bar and 10.5bar discharge pressures.

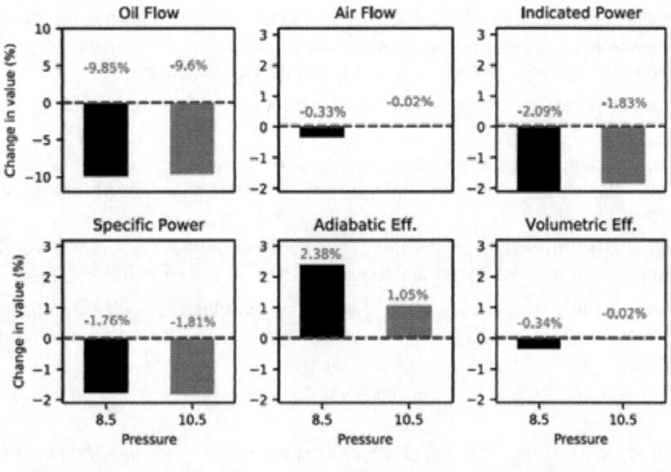

Figure 9: Comparison of the two injection ports and the single oil-injection port configurations at discharge pressures of 8.5 and 10.5bar. The green coloured text indicates improvement levels with the two port configuration

Conclusion

CFD analysis of an oil-injected compressor operating at 6000 rpm, and 8.5 bar and 10.5 discharge pressure is explored when oil is administered through single-point injection on the female rotor side. Key conclusions from the study are:

- Both power and flow predictions predicted through the CFD model were close to experimental results with differences of 0.5% for flow and 3% for power, respectively, at 8.5bar and 0.7% for flow and 2.3% for power at 10.5bar.
- This analysis method enabled both the oil and temperature distribution to be evaluated within the compression chamber.
- By increasing the injection port diameter, the discharge temperature is seen to be reduced, but beyond the optimal value of oil flow rate, no significant amount of cooling was formed. Therefore, it was decided to determine whether the maximum surface temperatures could be reduced by oil injection into both the male and female rotor sides of the casing.
- With two injection ports and the same total oil injection flow rate, the predicted indicated power was reduced by 2.1% and 1.8% at 8.5bar and 10.5bar, respectively. The specific power was reduced by 1.8% in

both cases, while the maximum internal temperature with the oil injection holes on both rotor sides was reduced to 127°C, which was an improvement of 30 – 35°C.

Techniques used in this study can be applied to a wide range of screw machines including refrigerant compressors provided suitable assumptions or developments are made to the VOF model. Also, similar to this study, a large number of performance results can be obtained through a combination of CFD and experiments for oil injection conditions such as oil flow rate, the number of injections points and injection positions at various operating conditions and compressor sizes. This data can be used to build a machine learning-based surrogate model to obtain the most optimal injection condition for a compressor size and operating condition.

References

[1] D. Vittorini and R. Cipollone, *Energy Saving Potential in Existing Industrial Compressors*, Energy, vol. 102, pp. 502–515, 2016, doi: 10.1016/j.energy.2016.02.115

[2] *Compressed air equipment presentation by Department for Business, Energy and Industrial Strategy*, 2020. https://assets.publishing.service.gov.uk/government/uploads/system/uploads/attachment_data/file/879754/Compressed_Air_Technology_Information_Leaflet_-_April_2020.pdf (accessed Oct. 03, 2020)

[3] S. Rane, A. Kovacevic, and N. Stosic, *CFD Analysis of Oil Flooded Twin Screw Compressors*, Int. Compress. Eng. Conf., no 2392, 2016

[4] R. Deipenwisch and K. Kauder, *Oil as a Design Parameter in Screw-Type Compressors: Oil Distribution and Power Losses Caused by Oil in the Working Chamber of a Screw-Type Compressor*, in International Conference on Compressor and their Systems, 1999, pp. 49–58

[5] P. Singh and J. Bowman, *Heat Transfer in Oil-Flooded Screw Compressors*, Int. Compress. Eng. Conf., pp. 135–153, 1986, [Online]. Available: http://docs.lib.purdue.edu/cgi/viewcontent.cgi?article=1520&context=icec

[6] N. Stosic, A. Kovacevic, K. Hanjalic, and L. Milutinovic, *Mathematical Modelling of the Oil Influence Upon the Working Cycle of Screw Compressors*, 1988, [Online]. Available: http://docs.lib.purdue.edu/cgi/viewcontent.cgi?article=1644&context=icec

[7] M. Fujiwara, *Performance Analysis of an Oil-Injected Screw Compressor and its Application*, vol. 18, no. 4, pp. 8–9, 1995

[8] N. Stošic, L. Milutinović, K. Hanjalić, and A. Kovačević, *Investigation of the Influence of Oil Injection Upon the Screw Compressor Working Process*, Int. J. Refrig., vol. 15, no. 4, pp. 206–220, 1992, doi: 10.1016/0140-7007(92)90051-U

[9] X. Peng, Z. Xing, T. Cui, and P. Shu, *Experimental Study of Oil Injection and Its Effect On Performance of Twin Screw Compressors*, nt. Compress. Eng. Conf., 2000

[10] M. De Paepe, W. Bogaert, and D. Mertens, *Cooling of Oil Injected Screw Compressors by Oil Atomisation*, vol. 25, pp. 2764–2779, 2005, doi: 10.1016/j.applthermaleng.2005.02.003

[11] A. Kovacevic, *Boundary Adaptation in Grid Generation for CFD Analysis of Screw Compressors*, Int. J. Numer. Methods Eng., vol. 64, no. 3, pp. 401–426, 2005, doi: 10.1002/nme.1376

[12] S. Rane and A. Kovacevic, *Algebraic Generation of Single Domain Computational Grid for Twin Screw Machines*. Part I . Implementation," Adv. Eng. Softw., vol. 107, pp. 38–50, 2017, doi: 10.1016/j.advengsoft.2017.02.003

[13] N. Basha, S. Rane, and A. Kovacevic, *Multiphase Flow Analysis in Oil-injected Twin Screw Compressor*, 3rd World Congress on Momentum, Heat and Mass Transfer, (2018), doi: 10.11159/icmfht18.132018

[14] N. Basha, A. Kovacevic, and S. Rane, *Analysis of Oil-Injected Twin-Screw Compressor with Multiphase Flow Models*, Designs, vol. 3, no. 4, p. 54, 2019, doi: 10.3390/designs3040054

[15] N. Basha, A. Kovacevic, and S. Rane, *User Defined Nodal Displacement of Numerical Mesh for Analysis of Screw Machines in FLUENT*, 2019, doi: 10.1088/1757-899X/604/1/012012

[16] *ANSYS Fluent User's Guide, 2019R1*, 2019

[17] N. Basha, A. Kovacevic, and S. Rane, *Numerical Investigation of Oil Injection in Screw Compressors*, Appl. Therm. Eng., 2021, doi: 10.1016/j.applthermaleng.2021.116959

[18] *ISO 1217:2009 Displacement Compressors — Acceptance Tests*, 2009. [Online]. Available: https://www.iso.org/obp/ui/#iso:std:iso:1217:en

[19] *Measurement of Fluid Flow in Closed Conduits*, 1987

[20] N. Basha, A. Kovacevic, N. Stosic, and I. Smith, *Effect of Oil-Injection on Twin Screw Compressor Performance*, IOP Conf. Ser. Mater. Sci. Eng., 2018, doi: 10.1088/1757-899X/425/1/012009

[21] *A Summary of Error Propagation*, Harvard University, 2013. http://ipl.physics.harvard.edu/wp-uploads/2013/03/PS3_Error_Propagation_sp13.pdf (accessed May 09, 2018)

Presented before the IOR on 3 February 2022

About the Author

Nausheen Basha

Nausheen Basha is a Chartered Engineer and Project Manager for a research programme aimed at advancing machine learning and computational fluid dynamics at Imperial's Chemical Engineering Department. She has a keen interest in fluid dynamics, multiphase flows, computing and improving diversity in engineering.

She graduated with a PhD from the City University of London, focussing on computational fluid dynamics modelling of oil-injected screw compressors.

DISCUSSION REPORT

Dermot Cotter, chairing thanked the author for an excellent paper.
It's strange, as refrigeration engineers, we look at compressors and we take them for granted. We assume you have some oil injection and we don't actually realise the core and the purpose of it.
I have a question about compressor motors and screw compressors - is the oil injected into screw compressors because you have an electric motor and you need to cool the electric motor? What is the real purpose of the oil injection into a typical type of refrigeration compressor?

Nausheen Basha
That's an interesting question. My research has mostly dealt with oil being injected into the air compressor, so obviously, in that case the important purpose of oil was cooling. We wanted to achieve something close to isothermal compression, so we needed to have the compression chamber cooled itself as the main purpose.

Dermot Cotter
And what happened to your screws if you didn't have oil cooling them? Would the rotor just seize up and lock or what would happen?

Nausheen Basha
Whilst designing the compressors, I was also designing compressors for a commercial client and there were instances when, if there was less oil in certain regions then the compressor would just stop working and you see the massive amount of heat and have to make sure that you switch off the whole system in time.

Dermot Cotter
Typically in refrigeration compressors, one part is oil for the rotors to make sure they don't seize up but also you have oil cooling in a lot of refrigerant as well which means it cools it down from a discharge of often around 80 or 90 degrees to around 55 degrees. Therefore you are actually cooling down the system and typically you don't have oil going through the motor, you actually have gas going through the motor to provide cooling.

A question from an on line listener - can you talk us through, in slightly more detail how the results were validated through experiments and how difficult the process to do it was with limited resources at the university?

Nausheen Basha
The big struggle for me to get the computer model running was stability. There were not many examples of resolving high pressure, oil injected compressors with very thin clearance gaps that have been achieved before. There have been some simulations but not to the extent of running a number of case studies. So getting the first thing right was quite hard and after that, once we knew the settings, and could run through things that was quite good. We had a level of validation that was done for the single oil injection and then we saw that quite closely the data matched with the results. I think somewhere around 2% which is quite a good level of comparison between experiment and theory. Once we had the computational model that was validated then we could use that model for testing a number of things.

Dermot Cotter
I have a question from Lars Anderson. Have the solid parts and the housing been taken into account regarding heat transfer, which besides direct impact on temperature distribution (perhaps not being

significant compared with the compression energy) can cause clearances to alter. Note, for decay on steel expands 100mm

Nausheen Basha
That's an interesting question as well and that brings us to another element in the model that could be considered going forward, where we could use more of a conjugate heat transfer being added to the model. So, currently when we look at the temperature distribution we look at this close to the rotor surfaces but it's not exactly rotor surface temperature. It's the gas temperature close to the rotor surface. So, I think that, in itself, answers that question. Going forward, and we have a layer of conjugate heat transfer with the metal that is being added, I think that would be much more realistic, but still, what this model tells us is how gas temperatures can be cooled and with gas temperatures cooling itself how efficiency can be improved.

Dermot Cotter
Can the intensive CFD used in future steps also, taking structural mechanic impact into consideration, reflect pressure mesh and bearing issues and elasticity, thermal expansion?

Nausheen Basha
It would be good. But I would say use CFD when it's needed and then make a sort of approximation within CFD rather than solving it as a whole system. Using CFD to understand temperature and distribution and then make a calculation on how the clearances and how the deflection would happen, how that would change. I think that would be an easier way than letting the CFD influence the structure to solve the problem.

Dermot Cotter
So, what you are saying is you need to come up with the idea first of all and CFD tells you the answer then you must reiterate and try another and if CFD doesn't come up with a solution it tells you if what you are proposing would work or not.

Nausheen Basha
It should be used in a way to understand if we make any design changes will it affect the results that we are looking for and then make the changes and then work out rather than having a CFD model to all sorts of flow structure interruption that would be too much for the model to handle.

Dermot Cotter
There is still a need to go back to the basics. You still need engineers when the computer doesn't tell you the answer. So, you focussed mainly on air compressors. I assume it was the air compressor because using refrigerants is more challenging and also, however the effects of using different refrigerants have different thermodynamic effects so do you foresee using different orifice sizes for different fluids?

Nausheen Basha
Yes, I feel this is actually the first step that gives us confidence to progress in that direction of understanding. How we can distribute the oil, how we can play with the oil injection parameters itself so, I think this gives us the confidence to go further then we could expand that to the refrigeration compressors and then understand how that would work.

Dermot Cotter
Have you considered, instead of having 2 ports having a port further up in the screw part of the machine that you might do an edit of oil further along the rotor rather than just at the start of the rotor. So in other words to try and have some at the beginning and some at the end so that the oil

that you had heated in a smaller quantity, if you used smaller quantities rather than two larger quantities.

Nausheen Basha
Yes, you mean one towards the suction side and another one somewhere towards the discharge side.

Dermot Cotter
Yes, a sort of economiser equivalent on the machine.

Nausheen Basha
Yes, I think initially when we designed the compressor, we thought to inject the oil at a certain temperature. So if we inject at a lower temperature which means that we want the heat exchanger to work in such a way that the temperature is much lower which basically adds to the load of the system itself. In our machine, for example we are injecting at 50 degrees temperature. We inject at that point of the discharge port where the temperature of the gas is higher. If we inject 50 degrees the suction itself, I think we are unnecessarily overheating the gears at the start of compression.

Dermot Cotter
When you tested the compressor with the two ports you were at 6000RPM. Did you look at reducing the speed or anything or did you see any effects in the compressor performances from reducing down the speed.

Nausheen Basha
We didn't change the speed so, one of the limitations is that we stayed at one speed. But I think, obviously, that speed has an impact on leaking the clearance gaps and also on the frictional losses the oil would experience and therefore power. We couldn't extend that to various speeds.

Dermot Cotter
So a screw compressor could be turned off or on. Thats because of the slippage and then the performance drops.

Nausheen Basha
Not really, no. What I'm saying is that this is an indicator that if you distribute well then there is a performance gain that is there to be made. This means that we could go in this direction a bit more and understand how the oil injection parameters and the speed are related as well and how we can use that knowledge in changing the design. I think more investigation is needed in that direction.

Dermot Cotter
I've another question from Lars Anderson – Have studies at Sevenca Rotor Machiner been reviewed and used as benchmarking? A lot of lab testing with oil and even water injection was performed and reported. Are you familiar with this?

Nausheen Basher
I'm not sure. SRM I guess that would be. I have read some of the studies and literature but not sure exactly about this one.

Dermot Cotter
Nausheen, it was a very good presentation and I have learned a lot today and I look forward to reading more of what you do in the future. Thank you very much.

IOR.org.uk
REFRIGERATION AIR CONDITIONING HEAT PUMPS

The knowledge hub for refrigeration, air conditioning and heat pumps

Ammonia Heat Pumps Helping to De-Carbonise Food and Beverage Industries

Robert Unsworth MinstR, Kenneth Hoffmann

Why you should attend

- A detailed case study of low carbon refrigeration at a major food and beverage processing site
- Learn how booster heat pumps can be used as part of a decarbonisation strategy
- Gain insights into the practical application and environmental benefits of large scale ammonia heat pumps

Abstract

In 2010 approximately 6.3 billion tonnes of food was produced globally where approximately 2.3 billion tonnes of this was via a refrigerated process. This has been increasing by approximately 10% year on year.

With approximately 17% of all power generated driving a refrigeration system of some sort, with an average coefficient of performance (COP) of 4, this means we are consuming 4,500 billion kWh per annum and rejecting approximately 18,000 billion kWh of heat. For the same heat, a boiler producing steam, burning natural gas (one of the cleaner fossil fuels) with a typical efficiency of 80%, would burn 22,500 billion kWh of fuel.

If we added a booster heat pump to all these systems and boosted the heat from +35°C (a typical condensing pressure of a refrigeration system in Europe) to +76°C (typical hot water loop temperature in a fresh milk pasteuriser, for example) our ammonia-based heat pump compressor could deliver this heat with a conservative coefficient of performance heating (COPH) of 6.5, thereby delivering 21,272 billion kWh potentially carbon free.

There are lots of generalisations above and obviously using all this heat is impractical if not impossible.However, the principal is still true. In the refrigeration industry, we are experts in heat (hot and cold). We are very good at removing heat, but also extremely good at throwing it away. Yet, heat pumps can help us to raise this heat to a useful industrial, commercial, and domestic process temperature, rather than it being rejected by condensers.

The vast majority of food and beverage production uses mechanical refrigeration to cool its process/products (plus factory space, people, lights and thermal losses etc) choosing to pay to reject this perfectly good heat via condensers, consuming more power, and possibly water and creating effluent in doing so. All because it's the "wrong temperature". Bizarrely, its often the same "heat/energy" which was added to the product, for example when pasteurising it, which is removed again by the refrigeration system. Just a different temperature. This heat is generally generated beside the refrigeration plant room, in the boiler house.

The following paper attempts to explain a little more how the right type of thinking can and is helping to decarbonise heat in food and beverage processes.

Keywords: Heat pumps, Decarbonisation, Compressors, Energy Efficiency, Optimised Design, innocent

Introduction

The first draft of this paper started from an assumption that the reader understood refrigeration and heat pump engineering. As a bit of a "geek" one's starting point tends to be from an assumption the reader is too. So, I've decided to start from the beginning, and explain what a heat pump is, how it works and how it's an important tool for decarbonising.

A heat pump is just a chiller with the temperature controller moved from the cold pipe on the evaporator to the hot pipe on the condenser (more or less) as it's the heat we are interested in and not how much cooling its giving.

Everyone has one at home, a refrigerator. When one buys product at the supermarket and the product warms up in the boot of the car on the way home, heat (or energy) is being added to it, as the laws of physics dictate heat travels from hot to cold.

This product, which is now warm, is put into the fridge, the heat passes into the evaporator in the back causing the refrigerant to evaporate at a low temperature. The heat causes the pressure in the evaporator to increase, the compressor sees this and starts to compress the evaporated vapour, keeping the pressure and therefore boiling temperature down.

The compressor, compresses the refrigerant to a pressure/temperature which it can condense, rejecting this heat to the ambient air, so again as hot travels to cold, the condensing pressure/temperature is hotter than the air in the kitchen. As a refrigeration cycle is a closed loop, all the energy/heat which entered the system in the evaporator (heat taken out the food) leaves at the condenser, but so does the energy which is driving the compressor. For example, if a refrigerator COP is 2.5, and it has a cooling capacity of 1kW then it will draw 0.4kW and the heat (which is hotter than the kitchen) will be 1.4kw. So, for 0.4kw of electricity you have now 1.4kW of heat warming the kitchen.

Open the door between the kitchen and garden and position the refrigerator in the doorway. Plug the spaces around it, open the door and turn it on, and now you have an air source heat pump, where the heat is being extracted from the air in the garden to heat the kitchen.

If we connect the condenser to the oven, and use the compressor to push the condensing pressure up higher, we would use, say another 0.4kw of power, but we would get 1.8kw out, but now we have heat we are using to cook with rather than just warming up the kitchen. If your oven is only 80% efficient, you would need to burn 2.25kw of gas to achieve the 1.8kw of heat you require, rather than an additional 0.4kw of electricity (remember you had to cool the food in the first place).

This is a simplified example of what we are trying to achieve in the food and beverage sectors in applying heat pumps. So rather than rejecting the heat at the condenser, increase its temperature and use it for something useful, replacing a fossil fuel based system.

The United Nations (UN) Intergovernmental Panel on Climate Change (IPCC) have announced that the world is in the middle of a climate emergency [1]. The climate change is real and human activities are the main cause. If the world is not able to keep the temperature rise below 1.5°C of pre-industrial level, we will see increasing sea water, dying coral reefs and danger to insects and vital pollinators habitat. In 2021 the main source of the temperature increase is the burning of fossil fuels which account to two thirds of the greenhouse gasses. Other sources are methane which accounts for 25% of the greenhouse gasses in 2021. Methane is 80 times more powerful than CO_2 (GWP20 = 80) when calculated on a 20 year life cycle [2]. In 2016 the members of UN Environment Programme (UNEP) agreed to the Kigali amendment of the Montreal Protocol on Substances that deplete the ozone layer. This amendment set out actions for reduction in production and consumption of hydrofluorocarbons (F-gasses) as they otherwise would lead to 0.4°C global warming and count for 25% of the global greenhouse gas emission by 2050. The effect on global warming potential (GWP) for F-gasses is based on their 100 years global effect. This has traditionally been a good guidance figure as the traditional F-gasses had a very long life cycle in the atmosphere, but many of the new F-gas refrigerants that have been introduced to the market since the Kigali amendment have a very short lifetime in the atmosphere, so it is more suitable to base their global warming effect based on a 20 years. Life cycles like it are already calculated for methane emissions for example. The new F-gasses have also shown to have other less favourable, environmental impacts. At one factory in the US they annually release R23 (which is a by-product of the HFO/HFC manufacturing process) with a global warming potential equal to 4,000,000 tons of CO_2. We don't know the emissions from all the other manufacturing facilities as they mainly are situated in less developed countries, so no one knows what the actual carbon footprint is for manufacturing F-gasses. There are also question marks regarding the breakdown components of the F-gasses which increasingly are detectable in our freshwater systems and are very similar to the fluoride salts (PFA's) which have been banned in many countries due to their toxicity to the environment.

In a race to reduce greenhouse gasses there is a major push to reduce the burning of fossil fuels. Fossil fuel is mainly used for heating, electricity generation and transport. Globally, heating amounts to 50% of the fossil fuel being used, and by optimising and decarbonising these heating systems large gains can be achieved to reduce the global emissions. One of the focus areas in decarbonisation is to decarbonise the electricity generation and electrify the heat production. To optimise heat production from renewable electricity sources, implementation of heat pumps would be an optimal solution. With coefficients of performance heating (COPH) between 3 – 10 (depending on the temperature lift) it is possible to gain 3 – 10 times the heat energy compared to the electric energy used.

Globally 19.3% of final energy usage is electricity (2018)[4], when we want to electrify heat generation by adding (what accounts for 50% of fossil fuel consumption) heat energy (heat pumps) to the electricity grid at the same time as transport energy consumption (cars, busses etc) is moving in the same way, there is going to be an enormous strain on the electricity grid in the coming years, which could potentially lead to interruptions of the electricity supply. The answer to decarbonisation is not just to electrify everything, it is very important to do it as efficiently as possible.

When we look at industrial heat pumps the difference between market average and best in-class can be more than 30%. It is important that the introduction of heat pumps into industry and district heating systems is done as efficiently as possible, not only to give the customer the best return on investment, but also to limit the impact on the supply infrastructure in the coming years. High temperature (in the area 60°C to 95°C) ammonia heat pumps have proven to be significantly more efficient than F-gas heat pumps and must play a major part in the coming years.

Compressor technologies for industrial heat pumps

Traditionally there has been three main compressor technologies used for industrial refrigeration: centrifugal compressors, screw compressors (single rotor or dual rotor design), and reciprocating compressors.

Figure 2: The picture shows three types of compressors typically used for heat pumps. To the left the multistage centrifugal compressor (Siemens Energy). In the middle a screw compressor (GEA) and to the right a reciprocating compressor (GEA).

As anyone involved in supplying refrigeration systems to food and beverage process understands, it is essential to consider load variations in systems to protect your refrigeration system and compressors. It would be nice if the design conditions of these systems remained constant and accurate, but I've not seen one that is yet.

The multistage centrifugal compressors are less capable of maintaining the pressure differential when the impeller speed reduces at part load. This leads to a drop off in efficiency and limits the minimum part load conditions. For positive displacement compressors the compressed gas is discharged at the same pressure independent of the rotation speed of the shaft and is just a function of your condenser performance (which is why it's important to keep your condenser in optimal condition, although as often as not, it is forgotten about as its away on the roof somewhere difficult to get at. Compressor designers argue that about 0.1% here and there in optimisation of the compression within the machine, but it is my experience most inefficiency's dwarf these figures before and after the vapour has entered or left the compressor).

Centrifugal compressors are more suited for moving large amounts of refrigeration gas which makes them suitable for F-gasses like R1234ze (R1234ze = 107 kJ/kg @ 82°C) as it requires ≥8 times increased refrigerant mass flow rate than ammonia (R717 = 865 kJ/kg @ 82°C) at the same heat pump conditions, also at fixed/stable loads.

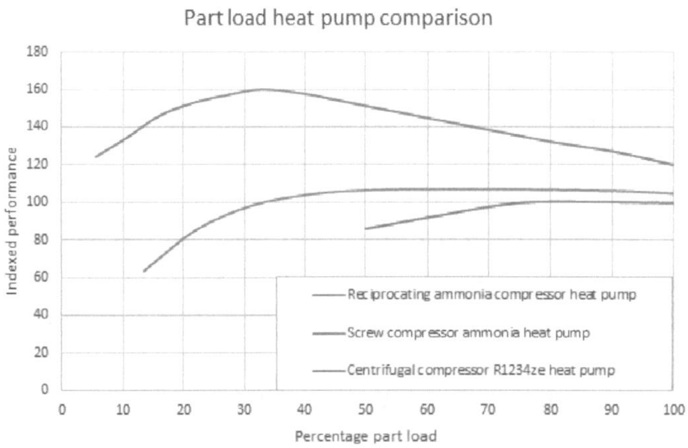

Figure 2: Part load comparison of three different heat pump technologies. The comparison is based on one project with index 100 for the efficiency of the centrifugal compressor at design conditions.

In the figure above, a comparison between the three different compression technologies where the heat pump's suction pressure is at 27°C and delivering hot water from 45°C to 75°C. This is just one operating point, and these curves will vary with varying conditions. However, one can see the efficiency of a screw compressor is closely matched to a centrifugal compressor at 100% load, but the screw outperforms the centrifugal at part-load.

The reciprocating compressor outperforms both rotating machines, and similarly to refrigeration this performance difference widens at part load (there is a reason we change gear in a car). The only limitation is that a high pressure reciprocating compressor for ammonia is limited to approximately 1000 m^3/h in comparison with screw compressors which are available up to 7000 m^3/h, so one would need more compressors with larger capacities which is why district heating schemes tend to opt for screw compressors, whereas most food and beverage facilities can be covered by a reciprocating compressor or two (especially when one considers the dynamic nature of the load over say district heating systems). The largest reciprocating compressor would be able to supply 2.7MW of hot water at the above temperature, which is approximately 4.5 tonnes of steam at the point of use, which is typically more than enough for food and beverage applications, especially when considering thermal requirements within their range <= +95°C with ammonia.

Incidentally, this temperature is more related to financial application than technical ability. This just happens to be the approximate temperature one can achieve when boosting from a condensing pressure of +35°C and keeping the spark gap and COPH within range of each other. Although even at +95°C, considerable carbon savings can be achieved, most users desire a reduction in OPEX too.

Heat recovery from ammonia refrigeration plants has been used for more than 100 years, via de-super-heating, subcooling practices. However, it is only in the last 15 years or so that dedicated high temperature ammonia heat pumps have been implemented into the industry and for district heating. Even these systems have typically served systems requiring lower temperatures, +65°C - +80°C due to the availability of components, expertise and return on investment.

In recent years the demand for higher temperature heat pumps in the industry has been increasing as it looks to decarbonise production. Investment in heat pumps must be realised in as efficient as possible manner to ensure power input, not only to reduce OPEX, but also to protect limited supply.

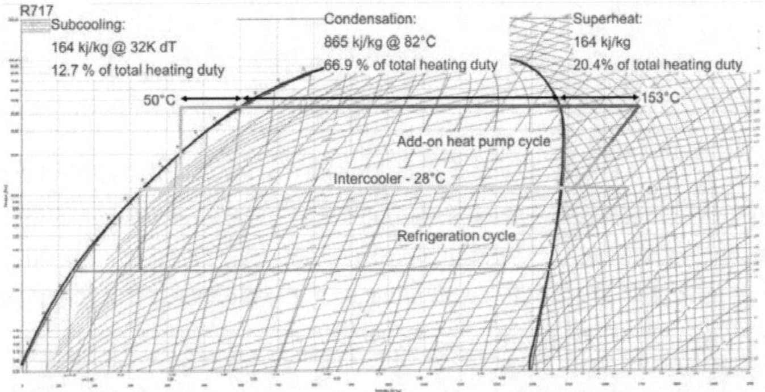

Figure 3: LogP-H diagram of a heat pump with reciprocating compressor recovering heat from the condenser circuit of a refrigeration plant.

Currently it can be seen that compressor ranges are being extended up to 95 °C for the different heating capacity requirements by the compressor manufacturers and existing first-generation products are replaced by new solutions.

Although there is a demand in most food and beverage factories for temperatures above 95°C, this is not cost effectively achieved as explained earlier, therefore other solutions should be considered. Although other refrigerants can be used to achieve higher temperatures, all have longevity or practical issues which shouldn't be ignored, however steam compression becomes more promising up to +200°C.

Steam can be created in a slight vacuum and then compressed to reach the required heating temperature quite efficiently. Mechanical Vapour Recompression (MVR) compressors of various types are already available, however they are dynamic compression machines and therefore have some limitations (apart from typically higher cost) such as differential pressure and maintaining discharge pressures at part load, which are an issue for most food and beverage systems because of the dynamic nature of their loads. That said, with latent enthalpies of steam being 2,600kj/kg and up, it is a perfect refrigerant for these elevated temperatures in a heat pump once a positive displacement compressor becomes available, and if created via an ammonia heat pump for example in a cascade arrangement, one could achieve these higher temperatures in a less costly, efficient and sustainable manner.

Once accomplished, this could mean zero carbon heating for 95% of all food and beverage production, in a cost effective manner.

An innocent application for heat

In the following section we will discuss the deployment of the above technology at innocent's new factory, the blender, in the port of Rotterdam. It is important when following the below information to think how many other applications this could be used in. For example, dairy, brewing, blanching, CIP, CQC, poultry and red meat, and good old wash down.

innocent wanted to build a sustainable factory and were looking at "any and all ways of delivering this". There were several consultants involved considering all sorts of energy sources, such as gas (natural and BIO) fired CHP combined with absorption cooling, among others. No one had considered heat pumps at the time because there was little refrigeration knowledge and understanding of the loads involved at that point in time.

Based on an expectation of tonnage of product (appx. 90,000kg/hr), it was possible to get an idea of thermal loads for both heating and cooling during the meeting (including room cooling in both production, offices and cold storage) and it was quickly discovered there was always an excess of heat leaving the refrigeration system than was required on the heating side of the factory and its processes.

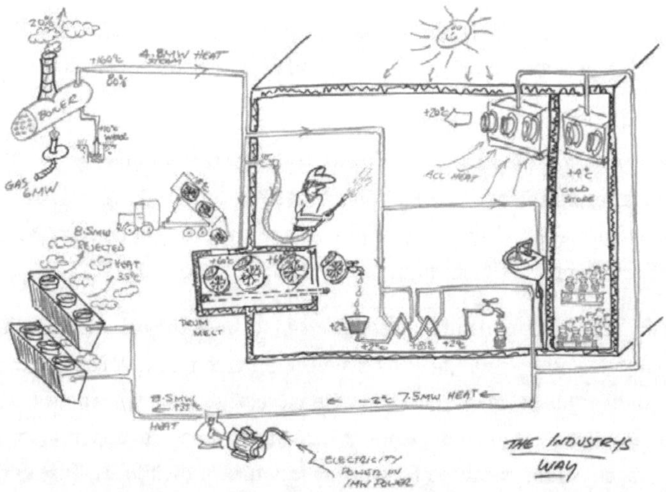

Figure 4: The original sketch of the "thermal performance" of the factory and its process using steam for all heating and refrigeration for cooling.

Above is the sketch I used in the meeting, attempting to explain the thermal process where a combination of chilled product enters the process at +2°C, but another stream is of barrels of frozen juice which is first defrosted in a tunnel with steam. Both temperature streams are stored in large vessels (250m^3) which are temperature controlled. The product passes down the various areas of the process, such as blending etc, and is thermally treated (pasteurised) at a temperature of 85–87°C.

Once pasteurised, the product is bottled in an aseptic filler, then stored in an automated chill store. The whole production area is "comfort cooled" because as you can imagine, although there are not so many people in these areas, there is a lot of automated equipment rejecting heat.

A pasteuriser works by taking in chilled product, heating to the required pasteurising temperature (the temperature of which depends on the product's chemical composition). For example, fresh milk might be +72°C, cream +78°C, juice +85°C, beer mid 70's etc. The product is held at this temperature in the pasteuriser for a time period (again depending on temperature and product type), typically just seconds. UHT treated milk is heated to higher temperatures but only for less than a second in some cases.

Tubular or plate type pasteurisers can be used, but this depends on product. In a brewery, tunnel type pasteurisers might be used in the case of canned and barrelled products where the product is pasteurised in its final container, but the thermal principal is the same.

The product enters and leaves at the same temperature in most cases, and the entering product is heated by the leaving product, minus thermal/efficiency losses. Heat is added via steam (traditionally) to compensate for the thermal losses in the process and these kW of heat must be removed again in the final stage of cooling within the pasteuriser.

Depending on the product type, and type of pasteuriser (which is again dependant on product type) will determine the thermal efficiency, but these tend to be between 75%-93% where a lower viscosity product like milk might be at the higher end, and a smoothie for instance, at the lower end.

Heating is normally performed by a closed water loop, controlled at an accurate temperature, via a steam injected calorifier. This is undertaken in this way to remove the risk of tainting the product by direct application of superheated steam in the heat exchanger where the product is on the other side.

The heat removed, via glycol for example in the final cooling section, is the same heat which was introduced into the product by the steam. It's the same kW, just at a different temperature.

In this example, the pasteuriser runs at an efficiency of 89% for juice (75% for smoothie), therefore at a product rate of 28,000kg/hr in a single line, this equates to:

(28,000kg/hr x 83k x 0.87kJ/kg.K)/860 x 0.11 = 258kW

If this was a smoothie product the flow rate would be reduced to around 20,000kg/hr, but the efficiency would decrease due to regeneration, so the thermal load would look like:

(20,000kg/hr x 83k x 0.88kJ/kg.K)/860 x 0.25 = 424kw

The type of product has a large impact on the equipment's performance and therefore energy consumption for the same temperature rise.

This would be more alarming when looking into a traditional method of heating and cooling because, we would need to supply 424kW of steam via a traditional boiler, and 424kW of cooling via refrigeration, therefore the overall effect would be:

434kW of steam via an 80% efficient boiler driven by natural gas consumes 542kW of fuel and producing 123kg/hr of CO_2.

434kW of cooling via a reciprocating compressor would have a COP of 4.4 and therefore consume 98.6kW (ET -7°C, CT +35°C). Add 15kW for condenser fans, glycol pump etc and your COPs for cooling ends up at a little over 3.8. We can assume at this point that the electrical power is supplied sustainably so the carbon produced is zero.

Therefore 545kW of burned fuel, 113kW of electrical power and 123kg/hr of CO_2.

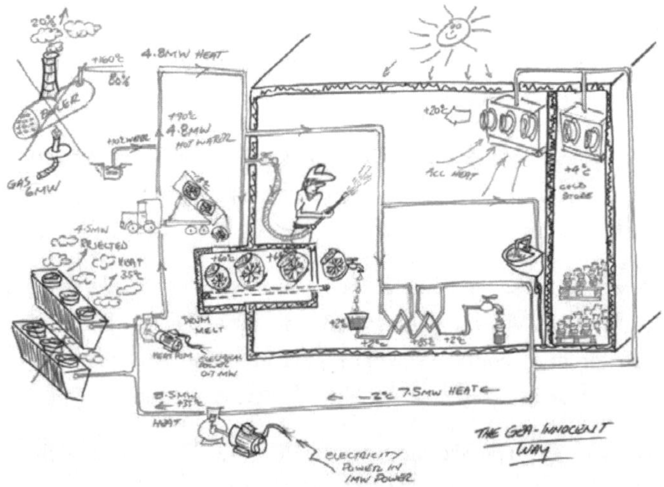

Figure 5: The original sketch of the "thermal performance" of the factory and its process using a simplified booster heat pump rather than steam.

Understanding the thermal loads

Typically, in a food and beverage factory, the boiler is controlled to produce steam at a given pressure. Although the user may only be heating water to +65°C in one process, if they are sterilizing in another at a temperature of 130°C, the boiler would operate at the higher pressure/temperature requirement irrespective of the capacity split. With losses in distribution and control, a boiler may run at a steam pressure of 6 barg or more (saturated pressure of 165°C).

In the case of innocent (and any other food and beverage facility) their heating and cooling capacities appear at different temperature levels, so for example:

-3°C: 5,000kW Cooling capacity. Process users including pasteurisers, product storage and cold store.

+7°C: 2,970kW Cooling capacity. Process conditioning, HVAC, and process cooling.

<=+65°C: 2,250kW Heating capacity. HVAC, hygiene, and barrel melting.

<=+85°C: 2,500kW Heating capacity. Process users including pasteurisation, CIP etc.

>+85°C: 2,300kW Heat via steam. Aseptic tanks and bottling.

Due to the capacity split it was decided to deliver a two temperature (-6°C and +4°C) suction system with the low temperature (LT) liquid being fed from the high temperature (HT) drum, and use hybrid condensers which although can run wetted, were sized to run dry due to the fact that we were using the heat pumps to control the condensing pressure and not the condenser fans and ambient air temperature. So, in effect removing the majority of the THR via the heat pump compressors.

When looking at the compressor COPs for each system, with a condensing pressure of +35°C:

-6°C Suction pressure: Compressor COP = 5.16

+4°C Suction compressor: Compressor COP = 6.7

Therefore, we have a combined thermal heat of rejection at the condenser of just under 9,900kW.

Using reciprocating compressor-based heat pumps, we produce the hottest achievable heat requirement by delivering +90°C water and directly injecting this water into the local pasteuriser loops.

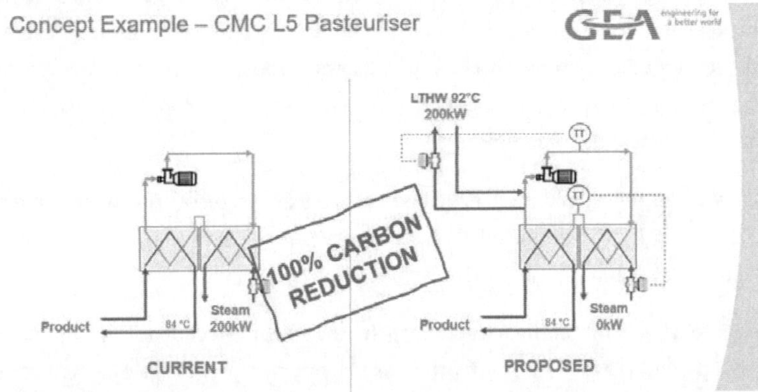

Figure 6: The standard method of heating a pasteuriser and the delivered method via a hot water loop driven by the heat pump

As we have used different oil types on the compressors of the high temperature heat pump than the lower temperature heat pump and cooling compressors, we use a cascade method of driving the high temperature heat pump off the medium temperature heat pump circuit. Doing so also helps with issues related to air purging as the systems are separated by a physical barrier at a higher pressure, so no non condensables enter the higher temperature heat pump.

The Compressor performances are as follows:

+90°C Discharge pressure: Compressor COPH = 8.0

Therefore the 2,300kW of Pasteurising load can be delivered for 287kW of power from +57°C

+65°C Discharge pressure: Compressor COPH = 8.1

The 2,250kW of heating load can be delivered for 277kW of power from the +35°C condensing pressure. This does however not take into account the additional power required to lift the pasteurising system load from the refrigeration systems condensing pressure to the suction pressure of the high temperature heat pump which absorbs an additional 248kW in doing so.

The total power consumed by the heat pump compressors to achieve the required 4,750kW of achievable heat (not steam) was 812kW excluding distribution pumps etc. If we calculate with 150kW of pump power to distribute the heat, we have a total of 968kW of electrical power, as opposed to 5,937kW of natural gas (NG) from heat delivered from steam in a traditional boiler, with an associated carbon emission of 1,347 kg/hr.

Large thermal buffer tanks were delivered for each temperature zone (cold and hot) for the secondary fluid. These compensate for the fluctuating nature of the cooling and heating loads in the process. It is important that large/sudden load fluctuations are buffered as to not effect any chiller or heat pump. A sudden load variation causing the suction pressure to drop, causes the saturated suction vapour to condense against the steel pipe and insides of the compressor as the pipe itself is now colder than the condensing condition, and therefore your suction pipe becomes a condenser. This phenomenon worsens with heat pumps due to the high suction pressure and therefore higher density of the vapour (a tonne of steel pipe, valves etc. 10k colder than the suction pressure will produce appx. 7ltrs of ammonia liquid to condense on the way to the compressor). Add to the fact that the suction condensing temperature is higher than the air outside the pipe. Where you might see condensation building up on the outside of the suction pipe due to poor insulation for example, or loss of vapour barrier in a refrigeration system, with heat pumps, the opposite occurs and the outside of the pipe remains fine, but the inside of the suction pipe becomes a condenser, and a compressor is a compressor not a pump. Always buffer to protect your compressor.

Other heat/energy and water saving initiatives have been achieved from recovering waste heat from compressed air to 80% reduction in water used via CIP systems, but I have tried to keep to information related to the application of the heat pumps in this paper.

One should also take into consideration the reduction of the total heat rejection (THR) by appx. 50% and therefore the power and utilities associated with otherwise having to reject this heat, when comparing. In fact, only approximately half the normal condensing capacity was installed because the process is continuous, and therefore so is the heat demand via the heat pumps.

In this instance, the final steam requirement is being delivered by a direct, electrically powered steam generator, and therefore the site has no fossil fuel supply connection at all.

All energy is delivered via sustainable sources (the site is located on the coast in the port of Rotterdam, directly opposite an extremely large offshore wind farm, and innocent are installing a pv farm on the site and will install their own wind turbines).

Even the location of the factory was chosen to limit the carbon associated with transport of raw and finished goods. In addition, working with local Dutch truck specialists on a 50T gross vehicle weight vehicle (a first in the market) some of the bulk juice which comes into the port of Rotterdam is now transported by battery electric hgv tanker trucks from the port to the factory where the trucks are re-charged while unloading product.

At the end of the day, all the above is perfectly fine, but solutions need to make commercial sense too.

Below is a ROI graph based on the above investment showing the return including savings in direct energy and carbon tax in the Netherlands.

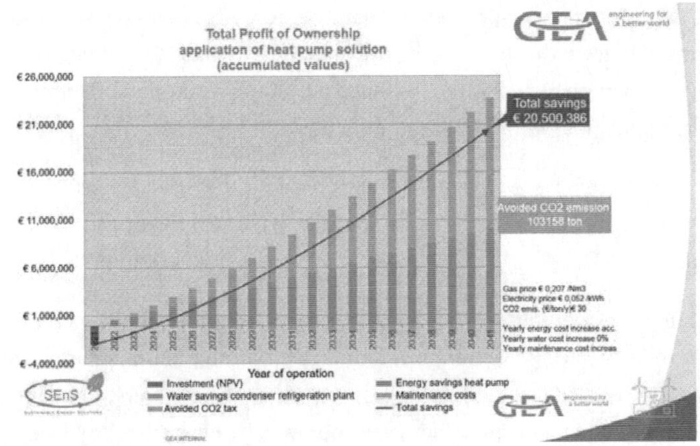

The above explanation is for a juice processing/blending system, but this is much the same as fresh milk in a dairy and to some degree soft drink manufacturing. Breweries can also benefit from this type of system, however the heating demands are typically greater than the refrigeration plant can deliver in this way (in Northern Europe anyway). A potential large cooling demand in a brewery is from cooling the in-coming water. Less of a problem in Scotland than Thailand for example. That said, even in the UK the refrigeration demand is enough to deliver heat to pasteurisers (for the same reason as the above example) and items like keg wash etc.

Chickens enter poultry primary processing plants quite happily at +38°C, and end up (less content) at either chilled temperatures of +2°C or frozen at -20°C. Therefore, there are large refrigeration loads in play, combined with process room cooling and cold storage, there is again a significant availability of heat at the condenser.

Typically, 0.6ltr of +52°C water/bird is consumed in pluckers and scalders in the process before chilling occurs. A line of 12,000 birds per hour would therefore consume 129.6 m^3 of hot water over an 18-hour production day. This equates to a little over 6,300 kWh per day. This volume can typically be doubled by the hygiene team washing down the factory at night, and this water would typically be stored and used at +65°C and therefore give an additional thermal demand of appx. 8,300 kWh. We, as refrigeration people could deliver this heat with zero carbon content as opposed to 3,300kg/day via burning natural gas.

As this water ends up going to drain, subcooling can be applied to the heat pump which raises its performance even further and therefore the heat pump compressor efficiency (COPH) for delivering this heat could be very close to 10. The steam used to heat at these temperatures does not improve significantly from the example above, even though the heat demand temperature is cooler.

Then you have examples of vegetable/rice/pasta production in CQC's (cook, quench, chill) where product is cooked/blanched in large vats of hot water before being quenched and chilled. Again, heating provided by steam and cooling provided by refrigerated chilled water, in a refrigerated factory.

Conclusion

Booster heat pumps have the potential to play a significant part in decarbonising the food and beverage industries.

All these industries require refrigeration systems and by adding the cooling capacity and compressor power consumption together, this is generally a good indication of how much heat is available to drive a booster heat pump solution.

Nine times out of ten, we as refrigeration people are rejecting more heat than the user needs (as long as they aren't flushing too much of it down the drain via the hygiene departments in their factories) due to the heat transfer to the product when its being processed, we need to remove again via refrigeration for the product to leave at its final/required temperature (frozen or chilled).

When we look round at processes, these types of solutions become glaringly obvious. The availability of higher temperature heat pumps has created more and more applications, which can provide efficient/sustainable heat where it is sorely needed.

As far as we are aware innocent were the first to take this step in decoupling their factory from consuming fossil fuels. Although we have been applying this type of technology for some years in areas of existing factories, where boilers were already installed and left "just in case", innocent were the first to go completely free of fossil fuels and should be commended for doing so. They really do stand by their ethos of "doing the right thing for the planet and people". innocent may have been the first, but they want to share the learnings, so that they are just the start and won't be the last.

We can envisage a day in the not-too-distant future, where the number of compressors entering the market for heating becomes similar to those used for cooling, as the world tries to decarbonise itself. One might ask oneself however, if this is indeed the case, rather than a Corgi registered engineer turning up to maintain your heating system, one now needs a Refrigeration Engineer, does the industry have enough of them?

References

Hoffman, K, 2022. *Ammonia Heat Pumps are Key to Global Decarbonisation*, DOI: 10.18462/iir.gl2022.0122

Joynson A, 2021, *Chief Blender*, innocent, paper presented at the IMechE Spring 2021

About the Author

Robert Unsworth

Robert Unsworth is the Technical Sales Director for GEA Refrigeration UK with over 28 years experience designing, installing, commissioning and troubleshooting industrial refrigeration and heat pump systems. Robert has worked on a range of projects including breweries in Thailand, freezer stores in the UK to juice factories in the Netherlands. He is also part of the new GEA SEnS Team which is an international team driving Sustainable Engineering Solutions for the GEA client base across all areas of its business.

DISCUSSION REPORT

Dermot Cotter, chairing, thanked the author for a very interesting paper.
The first question Robert is about what we mean by carbon neutral. For example, if I take out the gas boiler in my own house and put an electric heater in there and then I go to my electrical supplier and say can you give me green energy am I now being carbon neutral?

Robert Unsworth
You could do it another way – with photovoltaics on the roof and drive it with that.

Dermot Cotter
But if I don't and just say I am buying green electricity. In other words, I've made myself look green but I'm not really doing anything. If you use a heat pump you use electricity more effectively than using direct electric heat. An example, your 140-degree heat is being provided with electricity, which is a good idea and hopefully in the future you can do that with a heat pump as technology changes. But for heat pumps what you are actually doing is just increasing the efficiency of the electricity you are buying and converting to heat.

Robert Unsworth
At the end of the day, what you could do, or what the client could have done, would be just to put some massive electric boilers in and big windmills. The heating demand was I think, 4.8 megawatts plus half a megawatt for the steam for example. So, you would generate steam with electricity and then it would be cheaper to distribute the steam as you would with a normal boiler. Just generate it electrically, and it would have the same carbon footprint but less efficiency with regards to electricity. That then leads to, have we got enough electricity? So, as we all start using more and more electricity can the grid provide that amount of power? It's wasteful to not use it wisely. It's wasteful to use anything unwisely, isn't it?

Dermot Cotter
And you've kind of hit the nail with your subject topic there. You want to not waste electricity, and you decide to buy carbon neutral but the person down the road isn't because they just buy normal electricity.

Robert Unsworth
Efficiency is key. It's not efficient to keep throwing perfectly good heat away (condenser) just because it's not quite at the right temperature.

Dermot Cotter
I agree.

Robert Unsworth
In the example I showed we could have just gone for 90 degrees like you would with a boiler. Put a 90-degree heat pump in. It would have been cheaper to generate all the heat at 90 degrees, but you would have used more power to do it and you would still be "carbon neutral", however it would be wasteful.

Dermot Cotter
And it would cost you a lot more money to operate. So, really all this comes out to profitability at the end of the day because whilst you might use a boiler, putting in a heat pump is more profitable to the company, isn't it? And also, they could use the label of being green.

Robert Unsworth
Yes. If you take that on a bigger scale. If we all said we don't need to be efficient anymore, we'll just use electricity. But there's not going to be enough. Being efficient, it's the same with everything. Refrigeration is a massive user of power. It would be morally corrupt not to design our refrigeration systems as efficiently as we can.

Dermot Cotter
It all makes sense. I have a question from William Thompson. In 1983 a paper was presented to the BHRA on two heat pumps placed back-to-back for pasteurising processes at the milk marketing board site in Bamborough Bridge 40 years ago. Your application is identical, how do you think your process differs from this process and what additional energy saving does your modern solution provide to the dairy industry?

Robert Unsworth
I don't know that project, but ammonia is the most efficient refrigerant, so again, we shouldn't be wasteful. Ammonia supports that. I would never have picked another refrigerant for a job of this size other than Ammonia - it wouldn't make any sense at all. If you just follow the whole ethos of the client (innocent) it's not like they just want to be seen to be green, they actually really believe it and it's part of their being. So why would you use a refrigerant that's not particularly good for the environment when you've got a perfectly good one that is more efficient, and it doesn't have any side effects on the environment, so it just made sense. But, with regards to the specific project I don't know really because I don't know anything about that project. We put a heat pump in at a dairy 15 odd years ago and the difference here is the factory has been built with no connection to fossil fuel and that's a first. The pasteurization temperature of juice and smoothies is higher than fresh milk also, so the limitations in the 80's would have been the limit of compressor technology then.

Dermot Cotter
It's the right way to go, you need to do it from the design phase and it's very good.

Robert Unsworth
That's one of the issues. Typically, refrigeration experts might get involved in a project a year or even longer after a project has been started. This is much longer down the process than, for instance, my process colleagues, so the client has already had an architect and they have planned on building a boiler house and they might have already bought a boiler by the time it gets round to us talking about heat pumps. So, as an industry, I think the more we promote heat pumps and the efficient generation of heat and getting into these conversations earlier the better. So, it's in the mindset of the client. I think as people get used to heat pumps at home, when you go to work, and it becomes part of a project that you are working on then you will know what a heat pump is. I think it is just getting an understanding of what it is and how it can help you.

Dermot Cotter
So, for temperature requirements over 100 degrees centigrade do you also look at mechanical vapour compression? Is it there yet? How far away do you think is it?

Robert Unsworth
We use it already, so there's different companies around that use this already. Part of GEA uses it, we use pillar compressors. The only issue is it's a dynamic displacement machine, so it struggles working in a vacuum, it struggles with differential pressures, so you have to do it in multiple stages, which is why it is called multiple compression. To get up to 130/140 degrees you end up with a number of compressors in series. It gets quite complicated and quite expensive. We actually looked

at it at innocent but it was expensive to deliver a megawatt worth of heat so it didn't make financial sense. There are various companies looking at compression of steam in a positive displacement machine which enables you to raise the pressure with a single stage, so it becomes more cost effective. But the problem is those loads in food and beverage plants and dairies are so variable, which doesn't work very well with mechanical vapour compression but with a positive displacement machine we are used to variable loads. So, you maintain the discharge pressure even when it is running at part load. I think that's where the innovation is going to come over the next 5 years.

Dermot Cotter
Is there anybody on the market that finances retrofitting to existing sites?

Robert Unsworth
I know there's various energy companies, that sponsor finance projects for sustainability projects. We were approached a couple of years ago by a pension scheme - pension schemes have to invest a certain percentage of their funds into sustainable projects, so they are actively looking for financing projects. I in Denmark I put a heat pump in back in the 90s but that was district heating. In the last 15 years with the food and beverage side is where we've been applying heat pumps more and more. But they have always been "add ons" and the problem with heat pumps is that the heat pump might cost £200,000/ £250,000 but you have got to distribute the heat around the factory so you have to run a kilometre of pipe where they have already got a steam line in, so it ends up being that portion of the scheme that is cost prohibitive, actually distributing the heat. You may have to change a heat exchanger, so it's not the heat pump itself that costs the money. I think the benefit with innocent was that they had a green field site so they could decide to put a steam line in or decide to put a hot water line in and they chose a hot water line and the plumbing pretty much costs the same. So green field projects are better, more cost effective, however by no means a must. Now, whether you can get financing, or let's say the government help to finance, BEIS are financing heat pump projects to help with the burden of the additional cost of distributing the heat. There are finance schemes out there for adding heat pumps to existing systems.

Dermot Cotter
So, there is a question here. Are your heat pumps made in the UK?

Robert Unsworth
No. They are made in the Netherlands and in Germany. If we are building at a site-based plant like innocent then they are made wherever we build the factory.

Dermot Cotter
There is also a question about what kind of working fluids have you used for higher temperature applications and, about 150 degrees centigrade heat and the heat losses around the compressor bodies making the machines potentially inefficient. How are you overcoming these kinds of issues?

Robert Unsworth
We have got some experts looking at those temperatures at the moment. And, again, it is using steam. Kenneth Hoffman for example he has been working with this for some time.

Based on what we have been doing over the last few years using CO_2 to get up to 130 degrees, and with hydrocarbons, I just think it's limiting. Then obviously HFOs can be used for higher temperatures. Steam comes with a positive displacement machine, because of the benefits of steam. If you compress the steam, you only need the steam compressor and the heat pump, and you can distribute it in the existing mains, so you don't need to be running in lots of hot water pipework. You can use the existing distribution system, which is there and the existing heat exchangers so you would need a heat pump to get the heat up to 90 degrees first and then break into the existing distribution system. I think that is going to be the way heat 150-degree heat is going to be delivered in the future personally.

Dermot Cotter
So the low hanging fruit is up to 95 degrees centigrade and going above you are dealing with specialist areas.

Robert Unsworth
Yes, and I wouldn't go there if you are just getting into it now. But there's a lot of work to be done below 95 degrees. If you go around any food factory, the hundreds of cubic metres of water that they spray around the factory at night-time, consider that alone. Even if you are feeding a boiler with pre-heated water off a heat pump, there's just endless applications.

Dermot Cotter
In your experience, how many users have actually installed heat pumps where you actually don't have boiler back-ups?

Robert Unsworth
We've got 3 that I'm aware of, which have been installed over the last couple of years and that's predominantly because they are green field factories. The existing factories, they have got the boiler already.

Dermot Cotter
And once you have it you don't want to take that risk.

Robert Unsworth
Yes, so the example I showed you with the poultry plant. They installed the heat pump, 12 years ago and it's a piston compressor so it needs maintaining every third year or so. So, what tends to happen is you do the maintenance on the compressor and then make sure you get it back online quickly. They have got the existing boiler so that takes over the heating load, but they want the compressor up and running because of the savings. On that project alone, they were quoting £300,000 a year saving and it only cost them £500,000 to put it in originally.

Dermot Cotter
You talked earlier about the thermal storage tanks. How do you adjust the temperature stratification on them?

Robert Unsworth
We use stratification tanks these days. At innocent for example they are all stratification tanks, so the hot water is at the top and, obviously, the cooling water at the bottom. They are designed in that way.

Dermot Cotter
Thanked the author. I really enjoyed your presentation, and it has been a good question and answer session and you have shared with us a vision of the future, promoting technology, developing it, implementing it. Thank you very much.

Robert Unsworth

innocent are very keen to show what they do so if anybody wanted to go and have a look they would be more than welcome. As an industry I think it is important that we share this type of knowledge and get everybody into it.

IOR.org.uk
REFRIGERATION AIR CONDITIONING HEAT PUMPS

The knowledge hub for refrigeration, air conditioning and heat pumps

Carbon Dioxide Hazard Assessments and Safety System Requirements

Dr Dermot Cotter FInstR

Why you should attend

- A detailed review of hazards and risks associated with the safe operation of carbon dioxide refrigeration systems.
- An overview of the standards and other legislation to be taken into account for such systems.
- A template for a systemic approach to assessing of risk, emergency planning and facility awareness based on industry good practice.

Abstract

The safety of carbon dioxide refrigeration systems is a priority for owners, users and installers. Whilst legislation requires that refrigeration systems are to be designed to operate safely in all circumstances, the hazards associated with carbon dioxide are only likely to become apparent should an incident occur. EN378 is the design standard for refrigeration safety systems and includes particular requirements for carbon dioxide, within its text. EN378 details the circumstances for when gas detection, ventilation and automatic shut off valves are to be installed and their particular design requirements. However these safety systems are currently not well understood by many in the industry.

When carrying out a carbon dioxide hazard assessment, the outcome of the assessment is only as good as the knowledge of the assessor and their ability to identify the risks. A standard hazard assessment approach has been developed for assessing carbon dioxide refrigeration systems to ensure that relevant aspects of legislation and good practice are reviewed and recorded. Where non-compliances are identified, an action list is provided to guide the owner and users to implement changes to their refrigeration system, safety systems and working practices.

Keywords: Refrigeration, Carbon Dioxide, Safety, Hazard, Assessment

1. Introduction

In the UK, Parliament enacts legally binding Statutory Instruments (SI) and the Health and Safety Executive (HSE) acts as the regulatory body, responsible for its implementation. The SI's require equipment to be risk assessed by competent persons during its lifetime to ensure it is safe to operate. The publications related to carbon dioxide in particular are:

- *The Management of Health and Safety at Work Regulations*, 1999 (MHSWR) [1]
- *The Pressure System Safety Regulations,* 2000 (PSSR) [2]
- *The Dangerous Substances and Explosive Atmosphere Regulations*, 2002 (DSEAR) [3]

However while the legislation gives owners and users of carbon dioxide equipment legally binding responsibilities, there is little direct guidance in the regulations as to the exact requirements. The HSE publishes approved codes of practice for the regulations but this still leaves an element of expert interpretation on how the requirements of legislation can be implemented. There is however clear statements in regulations such as DSEAR that state that "Every employer shall ensure that risk is either eliminated or reduced so far as is reasonably practicable".

The European Committee for Standardisation (CEN) publishes standards for carbon dioxide refrigeration systems which support the essential requirements of EU Directives. In the UK the BSI has agreed that European standards are given the status of national standards and any conflicting national standards are withdrawn. Whilst other similar standards are written by the likes of trade bodies or associations, the European standard requirements should be implemented. The European standard publications related to carbon dioxide in particular are:

- EN378: *Refrigerating Systems and Heat Pumps – Safety and Environment Requirements* (Parts 1-4) [4]
- EN12693: *Refrigeration Systems and Heat Pumps – Safety and Environmental Requirements – Positive Displacement Refrigerant Compressors* [5]

Presented before the IOR on 7 April 2022

Whilst European standards are not legally binding, the safety requirements detailed in the standards are viewed by industry as "reasonably practicable" and therefore must be implemented to comply with regulations. These good standards can be used to assist the interpretation of how legislation compliance can be implemented and recognised by the courts.

Whilst the refrigeration standards focus mainly on the hardware safety systems that should be typically installed by refrigeration contractors, the owner and user have management safety system to implement to ensure their staff, contractors and visitors have adequate training, inductions and supervision.

The objective of this paper is to analyse the hazards when using carbon dioxide, review the safety systems requirements and how they should be practically implemented and provide an approach for preparation of a carbon dioxide hazard assessment.

2. Carbon dioxide characteristics

Carbon dioxide is an inorganic compound and has one carbon atom and two oxygen atoms. It has a modular weight of 44 and therefore as a refrigerant is designated as R744.

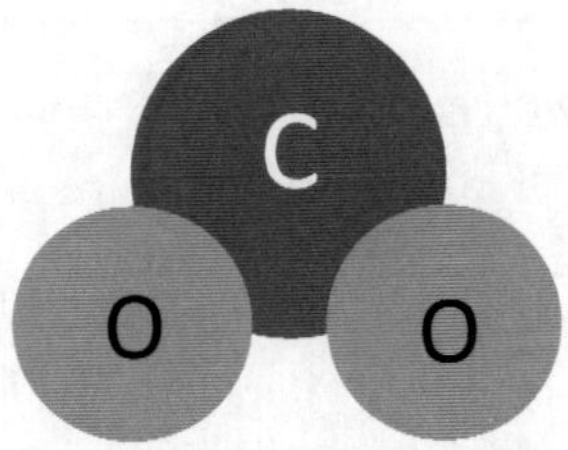

Figure 2.1: Carbon dioxide atoms

The European regulation (EC – 1272/2008) on *Classification, Labelling and Packaging of Substances and Mixture Regulations* (CLP) [6] classifies carbon dioxide, and it is to be labelled:

H280 Contains gas under pressure; may explode if heated.

H281 Contains refrigerated gas; may cause cryogenic burns or injury

H332 Harmful if inhaled

H335 May cause respiratory irritation

EN 378:2016, *Refrigeration Systems and Heat Pumps – Safety & Environmental Requirements*, classifies all refrigerants by safety groups, determined on their flammability and toxicity. EN 378, Part 1, Annexes E define the safety group for carbon dioxide as A1, meaning non flammable and lower toxicity.

2.1 Toxicity

Carbon dioxide is a colourless, odourless gas, denser than the air that occurs naturally in the earth's atmosphere. A person typically inhales and exhales air with carbon dioxide concertation of <0.05% and 4%. Whilst carbon dioxide toxicity is caused by displacing oxygen, lending to asphyxiation, the immediate danger to life would need to be in the order of 50% v/v. However carbon dioxide can cause an immediate threat to life at only 15% in air due to the toxicological impact it has on the body. The inhalation of carbon dioxide increases the acidity in the blood and thus trigger adverse effects on the respiratory, cardiovascular and central nervous systems. The reported effect of carbon dioxide exposure by inhalation (Public Health England) [7]:

2 to 5%	Headache, dizziness, sweating, shortness of breath
6 to 10%	Hyperventilation, tachycardia, worsening dizziness
11 to 17%	Drowsiness, muscle twitching, loss of consciousness
>17%	Convulsions, coma and death

Further analysis of the risk of fatality due to toxic carbon dioxide is given in Table 2.2 (HSE) [8]. It gives the likelihood of death due to exposure of carbon dioxide concentration with time:

Table 2.1: Inhalation exposure time versus carbon dioxide concentration

Inhalation Exposure Time	1-5% Fatalities Concentration in air v/v	50% Fatalities Concentration in air v/v
60 min	6.3%	8.4%
30 min	6.9%	9.2%
20 min	7.2%	9.6%
10 min	7.9%	10.5%
5 min	8.6%	11.5%
1 min	10.5%	14%

The UK HSE (EH40/2005) workplace exposure limit threshold limit value for an 8-hour period is 5000ppm, and for a 15-minute period, 15,000ppm.

2.2 Solid formation

When carbon dioxide reduces below 4.1barg (-56.1°C saturation pressure), dry ice (solid) can form through deposition. When carbon dioxide is released directly to atmosphere, the solid it creates will typically fall to the ground. When heat is added, the solid through sublimation will be released carbon dioxide gas without going through the liquid state. In the event of a release understanding the liquid and gaseous state inside the system will determine the quantity of dry ice created. Whilst gaseous carbon dioxide may only create 3% solid, liquid may create >30% solid by mass. The solid at atmospheric pressure has a temperature of –78.51°C.

2.3 Supercritical fluid

When carbon dioxide pressure rises to and above the supercritical pressure of 72.8 barg (31.04°C saturation temperature), the fluid exhibits the density of a liquid and viscosity of a gas.

2.4 Primary goal when designing CO_2 system

When any refrigerant is within a closed system and heat is added, its pressure will increase. In the UK the average ambient is 10°C. When a refrigeration system is turned off for a period of time, the low temperature side of the system will increase towards ambient temperature. With an average ambient temperature of 10°C, the resting pressure of a carbon dioxide refrigeration system would be 44barg. In the UK the ambient temperature is typically <32°C. For ambient <32°C, EN378-2 requires the low pressure side of heat exchangers exposed to the indoor ambient temperature to be designed at 27°C, this is equivalent to 66.3barg for carbon dioxide. For the high pressure side ambient, this would be 120barg. Whilst designers may make decisions and not fully follow EN378 guidance particularly for the low pressure side of the system, the lower the low pressure side design pressure, the more often refrigeration systems will vent carbon dioxide.

The priority of any risk based analysis would identify that keeping carbon dioxide releases to a minimum reduced risk. A refrigeration system that does not release carbon dioxide will not result in areas containing toxic gas atmospheres and therefore can be considered safe. Stopping all potential carbon dioxide releases is not possible and safety systems are still required, however the goal must be to keep potential carbon dioxide releases to a minimum. In the event of a carbon dioxide release the safety system must be adequate to keep this risk to a minimum.

3. Hardware safety systems for carbon dioxide

There are different requirements for safety systems for occupied spaces, machinery rooms and unoccupied spaces locations detailed in EN378.

3.1 Occupied spaces

EN378-3 section 4.4 states "refrigerating equipment located in the occupied space – the requirements shall be as specified in EN 378-1."

Detailed in EN378-1, the maximum quantity of carbon dioxide that is permitted in an occupied space is based on its access category and location of the refrigeration system components. For general access locations like supermarkets, hotels and dwellings, the charge limit is the toxicity limit x room volume. However with added special provisions like gas detection, ventilation and safety shut off valve systems the allowable charge of carbon dioxide can be increased. The same charge limit applies in supervised access and authorized access for locations in upper floor without emergency exits or below ground floor locations.

For all other access and locations areas, there is no charge restriction for the use of carbon dioxide in occupied areas. The carbon dioxide quantity could be 10kg or 1 million tons. It is only logical that special provision must be included for occupied spaces

Under section 7 of EN378-1 it states "where an indoor unit or related refrigerant–containing pipework, is located in a space such that the total charge exceeds the allowable charge, special provision shall be made to ensure at least an equivalent level of safety. See C.3." These special provisions are based on the room volume with the maximum floor area of 250m^2. When the refrigerant quantity exceeds 0.074kg/m^3 one special provision must be installed which can be gas detection with alarms, ventilation or safety shut off valves. When the refrigerant charge exceeds 0.18kg/m^3 gas detection with alarms must be installed and either ventilation or safety shut off valves must be installed. Sounders, beacons and alarm switches should be located within the occupied spaces.

Where installing gas detection is not practical in small chambers like spiral freezers, a detector just outside would indicate very high concentrations within a room. A risk assessment would also highlight the value in using personnel gas detection devices for measuring carbon dioxide concentrations when entering the room.

3.2 Machinery rooms

A machinery room is an unoccupied space without general access. The machinery room must be air tight from other rooms including roof voids. A housing which is sufficiently large for people to enter is a machinery room. Machinery rooms should be accessed by trained personnel and must have gas detection and ventilation systems.

The gas detection sensors must be located so if a leak occurs within the room, the extract fans will operate at full speed. The location of the air intake louvers and extract fan points must draw air across the carbon dioxide components reducing concentrations at any potential source of release. When a person enters the machinery room, the air flow in the room must be increased to a minimum of 4 air changes per hour for occupancy ventilation. It is important that the method of how the air flow is increased is taken into account. In machinery rooms with no external lighting, having the extract fans increasing in speed by turning on a light switch is the most reliable method. Using movement sensors with long delay off periods also works well when the sensors are located appropriately. Using signage is not appropriate for new installations due to the inherent laziness of people.

For machinery rooms, at least one carbon dioxide gas detection sensor should be located at the air extract point within the room and further gas sensors located based on a layout assessment. If the extract fan is not providing the four air changes per hour within the room for occupancy ventilation, the machinery room alarm must initiate. In the event of the carbon dioxide concentration reaching alarm levels, the extract fans should increase to the emergency air flow requirement based on the systems refrigerant charge or to 15 air changes per hour.

3.3 Opened air

Where carbon dioxide refrigeration equipment is located externally, there shall be force or natural ventilation. EN378-3 provides a definition of open air location. "A room, where at least one of the longer walls is open to the outside air by means of louvres with 75 % free area and covering at least 80 % of the wall area (or the equivalent if more than one wall is to outside), is considered as being in the open air." These open air locations must be assessed to determine if carbon dioxide can stagnate if released and gas detection and ventilation may be required.

3.4 Unoccupied spaces

Whilst EN378 gives a definition of occupied spaces, it considers the spaces above false ceiling and crawl spaces as forming part of the occupied space as the ceiling above is not air-tight. For roof voids above insulated chambers that are airtight to the chamber below, these spaces must be considered as unoccupied spaces even though in some instances the roof voids may be open at its edges to occupied spaces.

For non-permanent joints located in unoccupied spaces, EN378 section 4.5 gives the requirement for the space to be treated the same as a machinery room with air movements over the non-permanent joints. However where extract hoods are installed the air volume can be reduced down to >0.05m^3 based on the principle that the concentrated carbon dioxide is removed at source, and does not overspill the hood, thus removing the requirement for occupancy and emergency ventilation. The extract hoods must be designed taking into account the high likelihood of a carbon dioxide release occurs when a technician is working on the equipment inside the hood.

The design of the air extract hood is not detailed in EN378 and designers should not assume the extract hood is similar to that of a domestic canopy where warm steams mostly rises into the canopy and the extract fan function is to remove the steam from the canopy. Carbon dioxide is a chemical and therefore a chemical type extract hood must be installed that directs the carbon dioxide release to be extracted from the hood without spillage. It therefore should have roof, walls and floor. The roof can be a canopy, the walls can be overlapping strip curtains that allow easy maintenance access and a floor/drip tray to catch dry ice as per Figure 3.1. The air intakes can be a small gap between the tray/floor and wall. The extract hood must be smoke tested to check its efficiency. The overlapping strip curtains will not restrict access to the components and in the event of a carbon dioxide release, the strip curtains will fall back into position.

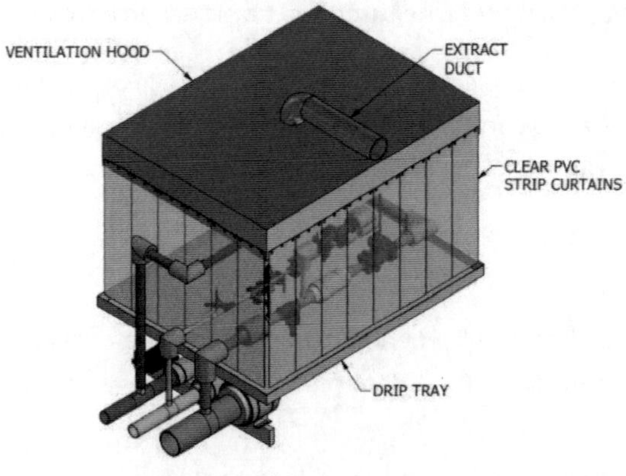

Figure 3.1: Extract hood

3.5 Relief valves

Andy Pearson et al (2021) [9], detailed the unpredictable nature when carbon dioxide relief valves vent and could result in a hazard with sublimation causing solid material to block relief valve vent lines. He recommends that further publications of EN378 include a note "the relief valve outlet pipe arrangement should be kept as simple as possible, avoiding branches, tees, sudden increases in cross section and other areas where solid material may deposit during a discharge. Preferably each relief valve should be piped individually to atmosphere with as few bends as possible."

A further consideration is the location and orientation of termination points for venting of relief valves. For larger carbon dioxide quantities where carbon dioxide is vented externally, the carbon dioxide must not be released in areas that will have little air movement resulting in areas that are above the maximum refrigeration concentration limits. Carbon dioxide venting horizontally may be considered lethal as any solids formed with exit at velocities typically >200m/s and may act like a bullet from a gun. Also any vented carbon dioxide must not be drawn back into buildings by ventilation systems, resulting in high concentrations within buildings.

4. Facility management systems for carbon dioxide

Carbon dioxide as with all refrigerants falls under DSEAR as it is a gas under pressure. Whilst the owner/user (employer) is required to carry out a risk assessment for carbon dioxide, this risk would be identical to that expected by the MHSWR regulations. However the risk assessment is only one aspect of the DSEAR regulations. The regulations also require arrangements to deal with accidents, incidents and emergencies. There is further requirement to provide employees with information, instruction and training.

Whilst the owner/user is responsible for compliance to DSEAR, EN378-2 requires the installer to provide information for safe operation of the refrigeration system and outlines the emergency procedures for the site to use in the event of a carbon dioxide release. This would highlight the need for training and having appropriate inductions on the site. The refrigeration contractor should not add carbon dioxide into the refrigeration system until all site management procedures are in place and refrigeration safety functional tests have been passed and documented.

The facility is also required to manage their contractors working on site.

4.1 Arrangements to deal with accidents, incidents and emergencies

A suitable and sufficient carbon dioxide release response plan must be in place. It must be well drilled, tested for efficiency and explained to staff as part of their induction when larger quantities are at the facility. The carbon dioxide release response plan should include the following:

- General information about carbon dioxide.
- Emergency contact numbers.
- A copy of the site induction that details where carbon dioxide is located and where staff, contractor and visitors shall congregate in an alarm. This carbon dioxide gas alarm may initiate the fire alarm to get personnel to go outside.
- Where carbon dioxide personal detectors are to be used, they must be regularly calibrated to the manufacturer's requirements and must be checked as part of any drill.
- A management structure and framework for responding to carbon dioxide releases (alarm and drowsiness/dizziness) and when to contact the emergency services and refrigeration contractor. A response flow chart is recommended – easy to follow in emergency situations.
- The response plan to include for leaks in all areas where carbon dioxide is present. This should include the occupied space and give detail of the refrigeration systems automatic responses along with those required by personnel.
- A package of information to be handed to the emergency services. This should include site layout drawing showing where carbon dioxide is present. The P&ID's of the carbon dioxide systems, carbon dioxide system quantity and COSHH sheets.

4.2 Provide employees information, instruction and training

All personnel working at the facility must have appropriate information, instruction and training. The facility induction must include:

- The location of carbon dioxide equipment.
- Understand the effects carbon dioxide has on the body and if concerned leave the location and inform management.
- Know the sound of a carbon dioxide alarm and immediate action to be taken on hearing alarm.
- What to do in the event of an alarm.

Where larger carbon dioxide quantities are located on a site, the management and those entering machinery rooms should attend a carbon dioxide awareness course.

4.3 Managing contractors

To reduce the likelihood of accident and incidences, the owner/user of the facility is required to have sufficient resources in place to manage contractors to HSE guidance HGS 159 [10]. When using a Permit To Work (PTW) system, those managing contractors must be competent to work to the HSE guidance HGS 250 [11].

Persons on site who supervise refrigeration contractors or those who work close to refrigeration equipment must have the skill to approve contractor's method statement, risk assessments, issue Permit to Works and supervision of contractors working on or close to the refrigeration system. Contractors may offer owner/users this training to those issuing Permit To Work to allow facilities to meet their legal responsibility.

5. Case study for new installation

A new carbon dioxide refrigeration system has been installed in the East of England. The carbon dioxide cools a ground floor (6m high) and first floor (6m high) chambers. The ground floor chamber is maintained at 2°C and has loading docks. The first floor chamber is only accessible from the ground floor and can either operate as a chill store at -2°C or a freezer store at -22°C. Next to this building is a 35m high freezer store operating on ammonia.

The carbon dioxide refrigeration system is a two stage system with two low stage compressors, four high stage compressors, vessels, five ground floor evaporators, four first floor evaporators and a gas cooler. The compressor, vessels and electrical panels are located within an enclosure on the roof located above the first floor chamber. The room evaporators and their valve stations are located within the occupied space.

Design

- Total carbon dioxide quantity: 550kg
- Quantity in pipe and evaporators: First floor: 160kg & ground floor 180kg
- Design pressure: Low stage 52barg & high stage 120barg
- Machinery room volume: 78m^3
- Relief venting: 2.35m on roof horizontally

Calculations

- Volume of each chamber (≤250m^2): 1500m^3
- 550kg in 1500m^3 space: 0.367kg/m^3 (exceed the 0.18kg/m^3)
- 180kg in 1500m^3 space: 0.12kg/m^3 (does not exceed 0.18kg/m^3)

From the calculations, it was identified that the carbon dioxide exceeds 0.18kg/m^3, therefore two special provision measures were installed.

- Carbon dioxide sensors with alarms in the occupied chambers.
- Automatic shut off valves within the carbon dioxide enclosure.

Two carbon dioxide gas sensors set at 5000ppm are located on each floor. When the gas sensor initiates an alarm, a sounder and beacon in the chamber will be initiated and the motorised valve on the main liquid and suction line to each floor will be closed. These motorized valves are located in the enclosure. This stops further carbon dioxide being released inside the occupied space that is stored in other parts of the refrigeration system. Non-return valves have been installed to allow excessive carbon dioxide pressure to bypass the motorised valves which will further reduce the refrigerant quantity released into the chamber.

The enclosure is classed as a machinery room and therefore occupancy and emergency ventilation has been installed. The machinery room is a walk in space, however the majority of carbon dioxide equipment can be accessed from the outside. A light switch is installed on the outside of the housing to force run the ventilations when the lights are on the housing. The fan provide 44 air changes per hour which is above the 4 air changes per hour for occupancy ventilation and 15 air changes per hour for emergency ventilation. Two gas sensors set to alarm at 5000ppm have been installed in the enclosure which will initiate sounders and beacons inside and outside the housing.

The relief valves are venting horizontally at 2.35m above the roof. The relief valves vent above peoples' heads. Carbon dioxide venting into the open area will not result in the gas stagnating.

The relief valves for the low pressure side are rated at 52 barg. This is equivalent to 17°C saturated suction pressure. Based on the EN378 5K difference between external equipment and internal equipment, the ambient would be required to rise to 22°C for an extended period before relief valves would vent. The ambient temperature is only above 22°C for less than 300hrs per year and the refrigeration equipment for this ground floor chill store will be running at this time. We therefore do not foresee, relief valves venting unless there is a power cut for a significant length of time and only if it occurs during the 3.5% of the year, when the ambient is above 22°C.

The management have vetted their refrigeration contractor and have received training from their installer on how to operate the refrigeration system. They have also updated their inductions and emergency release response plan adequately for carbon dioxide.

When the management on site issue a PTW to the refrigeration contractor, they verify the method statements follow safe working practices. These working practices for breaking into carbon dioxide systems include:

- Any valves opened or closed are listed on the method statement.
- Stopping occurrences of trapped liquid.
- Only vent carbon dioxide vapour. Liquid would solidify and potentially block service valves.
- Double isolation valves where possible.
- Wear appropriate PPE which includes goggles and chemical resistance gloves.
- Wearing a personnel carbon dioxide gas detector. Oxygen reduction detectors are not suitable.

From the method statement, the risk assessment would identify the carbon dioxide risks from its hazards have been reduced to acceptable levels.

6. Style and content of a carbon dioxide risk assessment document

The goal of any carbon dioxide risk assessment is to establish the risk rating (frequency x severity). Where the risk rating is found to be too high, the risk rating needs to be reduced by appropriate actions.

A standard format for carbon dioxide hazard assessments is a combination of a hazard study and risk assessment. The document is split into four sections:

1. Introduction, Actions and Compliance to Regulations
2. Carbon Dioxide Regulations, Standards and Good Practice
3. Compliance to Regulations and Good Practice
4. Risk Assessment

Section 1 introduces the document and then details the immediate actions required where the risk rating is considered too high under the headings: Pressure System Integrity Inspections, Site Management Systems, Safe Maintenance and Repair and Safety Hardware Compliance. There are further requirements related to the frequency the actions need to be carried out to ensure the refrigeration systems remain safe.

Section 2 identifies the limitation of the risk assessment, characteristics of carbon dioxide with COSHH and toxic effects. The section then goes into detail on: the requirement for avoidance of the release of a hazards substance, ventilation, the gas detection system, the safe release of carbon dioxide, carbon dioxide release response plans and on site personnel awareness requirement. This section would not normally change, as it covers all types of carbon dioxide systems, being based on regulations and standards.

Section 3 is unique and to be filled in for every site. It details the purpose and location of all carbon dioxide equipment, the gas detection system with auxiliaries, ventilation system safety systems, and details where carbon dioxide is vented directly or indirectly. There is then a list of compliance questions that assess if the system is safe and if it complies with the requirement as detailed in section 2. Where it does not meet the requirement, an action is labeled and then detailed in section 1.

Section 4 is the carbon dioxide risk assessment and establishes the risk rating for the present control measures and then future control measures when the actions are completed in section 1.

9. Conclusions

A systematic approach to assessing carbon dioxide refrigeration systems has been developed, to ensure any refrigeration system inspected has a logical and consistent approach for assessing and dealing with the risks identified. The approach to risk assessment is to prioritise the reduction of the likelihood of a carbon dioxide release and analyzing the effects of a carbon dioxide release with regard to toxic load. Further, in the event of a release, it considers if the emergency planning and facility awareness are suitable.

While regulations do not detail the safe working practice for refrigeration systems, European standards provide guidance of the safety requirement for carbon dioxide refrigeration systems.

References

1. Health and Safety Executive, 1999. *Management of Health and Safety at Work Regulations 1999 No. 3242*, HSE, United Kingdom
2. Health and Safety Executive, 2000. *The Pressure System Safety Regulations 2000 No. 128*, HSE, United Kingdom
3. Health and Safety Executive, 2002. *The Dangerous Substances and Explosive Regulations 2002 No.2776*, HSE, United Kingdom
4. European Committee for Standardization, 2016. BS EN378:2016 *Refrigeration System and Heat pumps – Safety and Environmental – Parts 1 to 4*, BSI, United Kingdom
5. European Committee for Standardization, 2008. BS EN12693:2008 *Refrigeration System and Heat pumps – Safety and Environmental – Positive displacement compressors*, BSI, United Kingdom
6. European Union, 2008. *European Regulations – Classification, Labelling and Packaging of Substances and Mixture Regulations (CLP)*, (EC – 1272/2008), Brussels
7. Public Health England, 2015. *Carbon Dioxide – Incident Management*, United Kingdom
8. Health and Safety Executive, 2011. *Assessment of Major Hazard Potential of Carbon Dioxide (CO_2)*, HSE, United Kingdom
9. A Pearson, Steve Benton, 2021. *Relief Valve Arrangements for CO_2 Refrigeration Systems*, Institute of Refrigeration, United Kingdom
10. Health and Safety Executive, 2011. HSG 259 *Managing contractors – A guide for employers*, HSE, United Kingdom
11. Health and Safety Executive, 2005. HSG 250 *Guidance on Permit-To-Work Systems*, HSE, United Kingdom

About the author

Dr Dermot Cotter

Dermot Cotter is the Managing Director of Star Technical Solutions the consultancy arm of Star Refrigeration Limited. In his secondary role as a Director in the Cold Chain Federation, he provides advice to its members on matters related to refrigeration systems and insulated envelopes. His interests and focus includes industrial and commercial equipment life cycle costs, optimisation of refrigeration systems, regulatory compliance and efficiency of low charge ammonia systems which was the focus of his PhD.

DISCUSSION REPORT

Colin Vines, chairing, thanked the author for an interesting paper. It's actually good to see, to really understand, how this reinforces the safety protection aspects of the CO_2 and the design of the system. It's not just about how can we stop costs of loss of refrigerant. Two questions. On a site with several CO_2 potential release points is there a simple formula or ratio that gives a good approximation of CO_2 release rates and the concentration from each point?

Dermot Cotter
You need to decide if it is a pinhole release or is a catastrophic release. Whilst there aren't many catastrophic releases you must prepare for the worst case scenario, particularly in an occupied space. So there might be numbers in standards like losing 10kg per hour but you must also think what happens if the joint fails. You might have the whole volume moving into the room in a very short space of time and that is why you consider the refrigerant quantity in the system. I haven't seen any analysis of what a flow pattern might be like in a room, but the reality is that wherever there is a release it builds up a concentration locally and then spreads out gradually.

Colin Vines
We have few questions around safety valves and relief valves. In the paper it says that you should only vent vapour, you should avoid venting liquid CO_2 because you will instantly create dry ice but then when you come to the sizing of the valve and the calculation of the sizing and lengths of the pipework, is there any guidance around that to prevent the potential for formulating dry ice?

Dermot Cotter
The reality is that you should try and put relief lines at the end point where possible so they vent directly to atmosphere and they shouldn't be vented into headers because you will always get little bits of crystals and you hope those crystals don't build up. The guidance coming from Andy Pearson's previous IOR paper has been to look and monitor very carefully and he gave no further advice but from my point of view you should avoid, where possible, having liquid lines unless there are special circumstances where you have no choice. It shouldn't be a typical supermarket pack. There may be unique applications where the charge is very small but the reality is you are putting the relief valves outside.

Colin Vines
Do you think retailers will apply your extract hood idea to the top of cabinets considering their costs?

Dermot Cotter
I never said it should be on top of cabinets. Remember it is an occupied space. If you are designing systems then you add in an extract fan more in the locations where there is a hazard so you are moving the CO_2out of those areas. It isn't all about fitting hoods because there may be a leak in an evaporator in a supermarket and it's a different environment. I'd suggest more localised ventilation and trying to build the design into the shop. Having your ventilation under the control of the gas detector as well. For the hoods, they are more for when the quantity gets larger. You definitely need gas detection. Remember if a gas detector goes off in a supermarket people don't know what it is and they don't move immediately. If it is a large leak then what might happen? That is where there would be a need to turn off the liquid supplies and take that into account if you are putting in very large systems. With smaller systems that hazard is reduced.

Colin Vines

How would you see these regulations applying to something like a bus with a CO_2 cooling system on it? Would the charge be very small?

Dermot Cotter
The first thing is to keep the charge down. There is no difference between a bus and an enclosed area. The big question is, is there ventilation on the bus? It isn't all about the charge quantities. If there isnt any fresh air coming in then you reduce the quantity of CO_2. There is a need for fresh air going through a bus full of people because you are dealing with levels of CO_2 from breathing. There should probably be at least 5 or 6 litres per minute per person if the bus was being treated like a building. I'm not an expert when it comes to transport and the air intakes.

Colin Vines
The paper you mention "Relief Valve Arrangements for CO2 Refrigeration Systems" by Andy Person and Steve Benton is available on the IOR website to download.
Is there any guidance on the frequency that leak detection systems should be calibrated or is that down to the manufacturers?

Dermot Cotter
First of all look at the manufacturers guidance. Some manufacturers say 6 months but typically a good quality gas detector should be annually. If it is lesser quality they might say 6 months.

Colin Vines
Is there an easy way of carrying out refrigeration risk assessments if the plant owner is not able to do it?

Dermot Cotter
If the owner of the system doesn't have a risk assessment in place then the likelihood of an incident really goes up because the management systems are not in place. It depends on where the site is but where there should be a permit to work. If a refrigeration contractor comes on site and they aren't analysing your risk assessments then automatically the likelihood of an incident goes up. Remember under HSE250 there must be resources in-house to manage the contractors or employ people to do so. Employers have a duty of care for staff and need to carry out an assessment to ensure staff are safe - but the reality is that most hazard assessments are done on behalf of the client and not on behalf of the refrigeration contractor.

Colin Vines
Is there a preferred method for CO_2 detection for refrigeration?

Dermot Cotter
It shouldn't be an aspirating type system. Any refrigeration detection system shouldn't be, for protection of people. It might be good for detecting low quantities of leaks but they're not good for toxic effects because of the delay. The reality is that when you are evaluating any system you are looking at it's response time. The response time should be less than 30 seconds. Once a manufacturer has proved that the detector does it's job and has an adequate response time I'm happy. Aspirating systems are good for detecting leaks but not for safety.

Colin Vines
Thank you Dermot. That has raised some interesting points.